The Amazing
Sea Otter

Victor B. Scheffer

The Amazing
Sea Otter

Illustrated by Gretchen Daiber

Charles Scribner's Sons | New York

Library of Congress Cataloging in Publication Data
Scheffer, Victor B.   The amazing sea otter.
Includes index.
1. Sea-otters.   I. Title.
QL737.C25S27   599.74′447   81–4557
ISBN 0–684–16878–2   AACR2

1 3 5 7 9 11 13 15 17 19   F/C   20 18 16 14 12 10 8 6 4 2

Printed in the United States of America

Contents

Address to the Beasts

For us who, from the moment
we are first worlded,
lapse into disarray,

who seldom know exactly
what we are up to,
and, as a rule, don't want to

what a joy to know,
even when we can't see or hear you,
that you are around. . . .

　　　　　　　—W. H. Auden

Preface

There is no other creature like it—it stands apart. It is Enhydra the sea otter.

It has always lived in the shallows of the North Pacific Ocean. Its habitat can be shown on a map by the thinnest of lines, for it thrives only in the narrow zone where submarine forests of kelp follow the contours of the shore. Deep down among the kelp stems, bathed in flickering light, it searches for food. When on the surface, it anchors its body with trailing fronds of kelp in preparation for sleep, and on a golden blanket of kelp it gives birth to its young.

It is a prime example of animal success. Long ago it abandoned the comfort and security of denning on land and elected (so to speak) to adopt a home forever cold and wet, open to winter gales and to the awful, sliding approach of predatory shark or killer whale. Yet Enhydra survived. It lives today as an encouragement to our own fumbling and uncertain species.

It is a marine mammal, sharing that distinction with whales, dolphins, porpoises, manatees, dugongs, and seals. In total more than a hundred species, the marine mammals

have a common past. All are descended from stocks that evolved in the sea, later adapted to living on land, and still later returned to the sea. They returned singly and gradually during 60 million years as though they were experimenting—which in a sense they were.

Enhydra, only 3 to 5 million years old, is the most recent repatriate to the sea. This is what scientists deduce from the fact that its fossil remains lie in the uppermost, or newest, rocky strata and that its living representatives, the modern sea otters, most closely resemble land mammals.

Enhydra is the heaviest member of the weasel family. Full-grown males from the California coast have been weighed at 86 pounds, females at 70. (Heavier ones have been weighed in Alaska—males and females at 102 and 72 pounds, respectively.)

Yet in spite of its weight, Enhydra is remarkably gymnastic. Twisting, rolling, and somersaulting in a continuous whirl of quicksilver motion, patting and smoothing its fur, it spends many hours of the day and night grooming its own luxuriant fur or that of its young. It squirms inside its loose skin like a cat in a bag. It seems at times to be wrestling with itself. It could well have been named the Acrobatic Otter.

It is the most social of the weasel-kind. When ancestral sea otters first entered the ocean they found themselves in a vast, open environment wherein quickly to establish social bonds of one kind or another promised survival to their species. Because modern otters often play among themselves, tease one another, and display something close to affection, they endear themselves to those of us who watch them. And because they assemble in large "rafts" they make life easier for those of us who study their habits.

Enhydra is the only mammal besides man and other primates known to use a tool. Much has been made of this point. I shall dwell on it later, now touching on it only as an example of the mysterious power of the ocean to engender what are some of the strangest of all animal habits.

And Enhydra alone among marine mammals has eyelashes. Does this trivial feature suggest how recently its ancestors lived on dry land? I think so. One who would understand evolution cannot ignore the small clues.

I first saw a living sea otter on the fourth of August, 1937, in the kelp beds of Ogliuga Island seven hundred

miles west of mainland Alaska. I had been tramping along
the shore of that bleak, windswept island at low tide, trying
simultaneously to watch my footing on slimy boulders and
to scan the kelp beds. Suddenly it dawned on me that a
certain ball bobbing a hundred yards offshore was *not* the
float of a kelp but the head of an animal. Through binoculars
it resolved into a staring, old-man face displaying a round
black nose framed by dripping white whiskers. . . . Then
it vanished, to surface a moment later. The animal had
spotted me and was moving downwind to catch my scent.
To my surprise, it swam on its *back,* twisting its head occasionally to choose its course. I watched it silently, enjoying the rare privilege of being one in tens of millions who
had seen a living sea otter, for in 1937 no aquarium had yet
displayed the species.

The setting of the present story is the central California coast along a reach of two hundred miles. Here the
swells of a great ocean slam against solid rock, or tumble
in confusion and lose their courage on sandy beaches, or
die quietly in submarine jungles of kelp. It was here in
March of 1938 that Frieda and Howard G. Sharpe, testing
a repaired telescope, "rediscovered" the California sea
otters. (A few had evidently been here all the time.) In
April of 1938 William L. Morgan of Monterey lowered
himself on a rope from a cliff to take the first known
photograph of California sea otters, a historic shot showing
about ninety animals floating together in a raft. When I
first looked at the photo I thought of dark chromosomes
strewn on a microscope slide.

This is the story of a sea otter called Barney, and it
is fiction based on fact. It tells of real and known events in

the lives of otters and of events that doubtless could have happened and could have been reported, had there been human witnesses. It tries to answer the question, *What might it be like to be a sea otter?* It describes sea otter anatomy and behavior in terms of their survival value to the species. What, for example, does an otter gain by having its big toe on the *outside* of its foot? By having a liver that seems to be twice too large for its body? By having lungs that can swell to a length of two feet?

The story also deals with people—Americans—for we Americans hold the fate of Enhydra in our hands. As daily we invade, exploit, and contaminate the waters where sea otters live, we face the question, *Is there room for us both?*

Except for the names of a few familiar regions such as San Francisco and Monterey Bay, the names of all places, persons, ships, organizations, and institutions in the story are fictional. Thus the research findings of several scientific organizations often appear—combined and condensed for the sake of readability—as the findings of one "Fishery Agency."

The principals in the story are

Barney, a newborn sea otter who is spared from living unseen and dying young when he is rescued by

Penny Moreno, a woman of eighteen years at the start of the story. She lives on the seacoast at Turtle Bay with her father (a commercial fisherman) and her brother.

Amelia Caring, librarian of the Marine Station. A pleasant, sandy-haired woman of middle age, she devotes her spare time to furthering the conservation of California's small, beleaguered sea otter colony.

And *Finn Peterson,* a biologist assigned by the Fishery

Agency to study sea otter ecology. Built as solidly as a mooring bit, he can wrestle a sea otter into a net or design a research plan with equal expertise. He is involved not only with otters but with sea urchins, clams and clam diggers, disgruntled lobstermen, television producers, divers, kelp harvesters, and various other life forms that one is apt to meet on the California coast.

My source materials are, by and large, the published reports of dedicated biologists in the United States Fish and Wildlife Service, the California Department of Fish and Game, the Alaska Department of Fish and Game, Sea World (San Diego), the Seattle Zoo, the Seattle Aquarium, the Tacoma Aquarium, the Vancouver (B.C.) Aquarium, the University of California, and Friends of the Sea Otter (Carmel). *The Otter Raft,* published twice yearly by Friends of the Sea Otter, is the only continuous, contemporary record of the California otter colony. Easily the best scientific book on Enhydra is Karl W. Kenyon's *The Sea Otter in the Eastern North Pacific Ocean* (1969). It was my thirty-year friendship with Karl, as well as our shared adventures in the haunts of sea otters, that prompted me to write the present story.

With deep feeling I dedicate the story to the late Betty S. Davis (1921–81), who critically read the manuscript.

The Amazing
Sea Otter

Prologue

Approaching the Year of the Sea Otter

When Penny awakens she is fully awake. The stillness of her room—a room in which the sounds of the Pacific Ocean are seldom absent—has broken her sleep at dawn. She smiles, lifts her brown arms over her head and thinks, *This is* my *day; it will be different.*

She pulls on faded blue jeans and a sweatshirt, ties a yarn through her shining hair, and hikes on morning dew toward the beach. Her clear warm skin, her dark eyes, and even the bounce of her hair as it moves with her stride, bespeak her heritage of Old California blood. In the east the sun is touching the dry hills above Turtle Bay; in the west a cottony fog bank presses upon the sea.

It is June. Penny is at home waiting to enter California Institute, where in the fall she will take up studies in engineering. From childhood she has enjoyed the lines and shapes of things, and the kinds of orderly patterns that engineers design and fashion into bridges and towers, tunnels

and machines. She sees beauty in nature, too, in the curve of the pelican's wing as it glides only inches above the wave, in the architecture of the animal skeleton lying upon the sand, and in the power of the living root to split the granite rock.

Now she strides along, squeezing the damp sand between her toes. She is alone; her footsteps are the first to call forth whispers from the crust of the beach. She hears birds piping and turns to watch a flock of sanderlings play tag with the waves, then lift in alarm and wheel in unison as though controlled by a single mind. She sees where during the night the fringe of the tide had felt and identified the size, weight, and shape of sand grains, pebbles, bits of wood, seaweed, shells, feathers, and foam.

Now as she stares toward the lightening hills her thoughts return to a problem of the spirit she has not been able to resolve. She speaks softly to herself.

"Should I really enter engineering? I'm not sure . . . It seems so awfully *tight* and *practical* . . . Whatever I do I've got to be free. I've got to be able to put aside blueprints and rules and exact ways of doing things. Engineering seems . . . so far from the *earth,* and I guess I'm an earth person. Should I switch to art? But how does one make a living at art? I don't want to be a starving artist. . . ." Her thoughts run ahead of her words.

Waah! Waah! A shrill call like the voice of a young gull in pain turns her head toward the sea. The call rises from a tangle of golden seaweed torn loose by some current that later coiled it upon the sand. As Penny nears the source of the sound she sees a yellow-brown kitten—or is it a dog? Its damp, silky wool is peppered with sand. Again it cries, and Penny thinks of the squall of a rubber doll she stepped on as a child. Her intuition tells her that the animal is young, helpless, and frightened. Stripping off her shirt, she wraps it around the baby, avoiding its teeth, and carries it on her breast toward home. Reassured by body contact and warmth, it stills its voice and its struggles.

When Penny unwraps her burden on the kitchen floor, she interrupts her brother at breakfast.

"Wow! Where'd you find the sea otter pup?"

So *that's* what it is.

"About five minutes walk from here, near the surf. How do you know it's a sea otter?"

"When I've been fishing along the kelp with Pop I've seen a few of them—but never this far south. They hang out near Rocky Point. *Now* what are you going to do? Did anyone see you take it? There's a big penalty for picking up even a dead one."

"Well," reflects Penny, "I'd better phone the game

warden and tell him I've rescued an abandoned sea otter. Then I'll call the veterinarian and ask him what to feed it. I'm sure he'll help. I'm certainly not going to put it back on the beach to be torn to bits by dogs and crows."

The warden who answers her call is at first annoyed that she has removed a protected species from the beach. He knows that often a well-intended beachcomber will "save" a young seal or sea lion whose mother is, in fact, only temporarily absent.

"Tell me again exactly where you found it. Maybe its mother is looking for it. I'll check the place and meet you later at the vet's."

The warden drives slowly along the wet beach toward the mound of kelp described by Penny. Almost there, his trained eye spots a cluster of gulls fussing over something in the tidal zone. He approaches on foot and sees a bird-spattered body. The empty eye sockets of an adult sea otter face toward the sky; two black nipples protrude from its belly. As he drags the carcass toward the surf to wash it off for a closer look, dark blood dribbles from a hole in the forehead. "A bullet," he mutters. Sickened and angry, he knows that the otter has been shot and that it is surely the mother of Penny's pup. He notes, in writing, the time and place of the find, loads the stiff body in his van, and drives away to put it in cold storage. Finn Peterson, the biologist, will want to autopsy it, for an otter having just given birth is seldom available for study.

Penny carries the pup, bundled in towels in the basket of her bike, to the veterinarian.

The vet's eyes soften as he holds the pup between his hands and stares into its expressive face. "I'll do my best to feed it, but I don't know the composition of sea otter milk and I doubt that it's ever been analyzed."

He picks a book from a shelf. "Here's a formulary of milk replacers for young zoo animals and pets. Hmmm . . . land otter . . . California sea lion . . ." He decides on a mixture of whipping cream, codliver oil, casein, salt, vitamins, and warm water, and, leaving Penny to swab down the pup, goes off to collect the ingredients.

As Penny gently towels the pup from snout to tail it trembles and tries to suck her hand. She studies its pudgy body covered with fine, cinnamon-brown wool and long, silky, yellow-tipped overhairs. The stub of its umbilical cord is still fresh and pink. Hefting the pup, she estimates its weight at four pounds. Its ears are thick and stubby and are nearly buried in the wool. Its hind feet are enormous and are webbed somewhat like those of a seal, while the longest of the five toes is on the outside where the shortest should be. (She will learn later that a sea otter spends a good part of its time upside down, backstroking the water with its flaring toes.)

"What a weird, lovely thing you are. Are you sure you haven't got your shoes on the wrong feet? I'll give you a name." She grasps the loose skin of its back and turns it belly-up on the towel. "A little male . . . I hereby christen you Barney." Her lips curve as she remembers out of the misty past another Barney, a fabric animal of uncertain species with whom she shared her blankets when she was six years old.

As she strokes the pup she soothes it with the little mother-words that in many mammalian tongues are similar. Barney takes a clumsy, experimental step on the floor. He stumbles and falls on his side. Clearly his muscles are not as well-developed as are his dense wool, his black-button eyes, and his teeth. He is a creature more of the sea than of the land, a creature designed by nature to float and dive rather than to walk.

His drying wool fluffs into a golden fleece in the morning sun. He is transformed.

The vet returns with the milk ingredients and mixes them in a stewpan. "I wonder how warm the formula should be." He turns to another handbook. "Hmmm . . . deep-body temperature of Alaskan sea otters, 38.1 degrees Celsius (100.6 degrees Fahrenheit). A bit warmer than the human body. Strange. I would have thought that a sea mammal would be colder than a land mammal. Still, I don't know why it should be."

Barney readily takes the artificial milk from a bottle and nipple. When he has swallowed six ounces the vet cautiously pulls the nipple from his mouth. "We'd better go easy the first time. We'll eventually learn his dietary needs and his natural rhythm of feeding. Anyway, I don't know how to *burp* an otter."

He lines a box with paper towels and puts the now sleeping Barney to bed. What an ordeal the little fellow has been through! For many hours—perhaps a day or two— he has suffered hunger, loneliness, cold, and fright.

"When he wakes up will you help me again?" asks the vet. "While I was in town I called Finn Peterson at the Fishery Agency and told him about our problem child. He advised us to give it a bath as soon as possible—said that a baby otter in the wild is more often wet than dry." (The vet did not realize that the pup had been *born* in water, on a seaweed mattress where he had been tickled into first breath by fronds of kelp.)

Agreeing, Penny returns to her home for breakfast. She comes back at noon to greet an infant who demands food in a shrill, whistling voice—a voice not quite a bawl nor yet a meow. Fed again on formula, he is lowered gently into a dog's bathtub filled with cool water.

Now he is clearly in his element. He floats high in the water on his back, holding his short arms, feet, head, and tail in the air. He rides as though his body were a bag of corks. Occasionally he pats his forepaws together, sneezes, and shakes water from his head like a dog.

"If we leave him alone, perhaps he'll clean himself."

But the vet is mistaken. A sea otter pup is wholly dependent for several weeks upon its mother. If Barney were now in the wild he would be lying on his mother's breast, enjoying her caresses. She would be using her tongue, paws, and warm breath as towel, brush, and dryer.

After ten minutes, Barney rolls over on his milk-swollen belly and, simultaneously putting his face under water and flailing the surface with his paws, tries to swim. His clownish attempt fails; he returns to floating position and relieves his bowels.

"Interesting," muses the vet. "I once helped the Fishery Agency livetrap a beaver. We held it in a pen where we fed it apples, carrots, and cottonwood limbs. It refused to defecate until we carried it to the river and let it sprawl naturally—then it let go."

The vet rinses Barney in clean water, dries him with a thick towel, and leaves the office to make his daily rounds.

In the weeks that followed, Barney grew increasingly playful after each meal. He loved to be rolled on the floor and to be prodded with Penny's fingertips like a beanbag.

His good spirits were evidence of health, yet his awkward movements and the slackness of his skin bespoke an ailing creature. Was he doomed to die as certainly as do most newborns cast as orphans upon the sand? Penny was encouraged by the vet to believe that he might pull through. Rarely now did he voice the piercing cries that earlier had announced his panic at finding himself alone and helpless in a strange place. He accepted a feeding schedule of four meals a day—at five-thirty, in the half-light of dawn, and again at ten, two-thirty, and seven. Had he been at sea, floating in his mother's arms, he would have nursed whenever he was hungry, at least six times a day.

Fortunately for his adoptive parents, he slept soundly through the nights. He paid no attention to the barking of dogs in the kennels behind the doctor's office, for his tiny brain had not been programmed to accept those sounds as danger signals. He often turned and twitched in his dreams, pawed the air, and nodded his head, rolling to the unfelt stir of an ocean he had barely known.

When the vet offered him a spoonful of fresh minced clam, he accepted it with relish, although perhaps only as a human baby would accept and enjoy for its texture a teething ring placed in its mouth. Then the vet had second, and more cautious, thoughts; he postponed further offers of solid food until Barney was four months old.

In the meanwhile, news of Barney's rescue has reached the headquarters of Otters Alive, a wildlife conservation group. Its members, numbering in the thousands, are dedicated to helping the California otters return to a population level in balance with the carrying capacity of the coastal waters. Otters Alive is a reservoir of kindness—and of collective expertise, for among its members are biologists, lawyers, physicians, divers, and photographers who apply their special talents to helping solve the dual problems of understanding sea otter biology and of understanding where sea otters fit into man's tight little world.

The "den" of Otters Alive is a remodeled sardine warehouse on Cannery Row in Monterey, donated by a man who made a modest fortune from harvesting kelp. The lovely walls of the building, now decorated by conservation posters, tell of a bygone era when fourteen-inch, straight-grain redwood boards could be had at any lumberyard. And on quiet nights the descendants of mice who knew Steinbeck scamper along the beams and sniff at cracks that carry to this day the oily traces of a billion fish.

So Otters Alive springs into action. It agrees to help raise Barney to the age where he can safely be returned to the ocean.

Penny and the vet, the surrogate parents, discuss what may happen when that time of parting arrives. Says she, "I'm worried that, having been raised by people, he will never *know* he's a sea otter. Not having learned by watching others of his kind, he won't be able to feed himself, or defend himself, or keep himself clean."

"What you're wondering," muses the vet, "is whether the conditioning we're imposing upon him will smother his instincts to behave as an otter ought—no pun intended. You may be right. I knew of a harbor seal that was held in an outdoor pool by himself. Because he never heard the voices of other seals, but did hear the barking of dogs attracted to his pool, he learned to bark like a dog!"

"My brother thinks we ought to turn Barney loose right away in a kelp bed somewhere north of here—put him back with other otters. He says it's wrong to keep him as a pet."

The vet smoothes his mustache. "I can't agree. If we succeed in raising him to independence, what we'll have learned will help others to care for other orphans." He frowns. "And the word 'pet' turns me off. I think of Barney simply as a bit of nature we've borrowed to study—and I hope to help—before we turn it back to nature."

"Well, when the day comes to turn him loose, all we can do is to say a little prayer. I'll work on the prayer . . . you keep him alive and well."

Otters Alive sends a young woman volunteer to live in a van near Penny's house and assigns her the task of sharing with the feeding, bathing, and exercising of Barney. Otters Alive will share costs of the volunteer's keep and the pup's food.

However, after several weeks have elapsed, the vet is disenchanted with the new arrangement. Such a coming and going of visitors to see the baby otter, and so many moppings of the floor, and so many dirty towels, bottles, and stewpans!

Now Finn Peterson intercedes. "I've been talking to my supervisor. We suggest that you move Barney to the Turtle Bay hatchery. It has clean, circulating seawater, a fish-food kitchen, and a small screened pool where Barney will be safe from dogs and vandals. Our agency will be gaining new and useful data on the early weeks of a sea otter pup. In the interest of science, we'll be glad to provide the facility if you'll cover Barney's food and medical care."

A splendid compromise. The vet gives the pup a precautionary shot of antibiotics. The volunteer moves her van to the hatchery grounds under the shade of a giant cypress. She anchors a sloping catwalk at the edge of the pool so that Barney can crawl in and out of the water.

When first introduced to the pool, Barney is confused by its shimmering expanse. He stumbles along its concrete edge, sniffing the ground as he goes, retreating often to rest on the volunteer's bare feet. Several weeks will pass before he will venture on his own to leave his poolside bed for the water.

In October, after helping the volunteer get Barney started on a new feeding schedule, Penny leaves Turtle Bay to enter California Institute. Now the pup will get chopped squid, shrimp, and clam in addition to drink.

He soon learns to feed while he floats in his pool. Bathing and eating at the same time proves so delightful that, while still sucking from the bottle, he struggles to leave the volunteer's grasp and crawl toward the water. She yields, and thereafter, clad in a bathing suit, she enters *his* element to feed him.

Wholly untaught, his behavior triggered by instinct, he begins to "clear the table" after each meal—that is, to clean his own chest and belly. He pats his pelage like a drummer, then makes corkscrew rolls to rinse all sides of his body.

"Barney, you're pure ham," says the volunteer, smiling, as she sees him practicing daily the body skills that later will keep him alive.

His four meals a day are advanced to the hours of seven, ten, four, and seven—times more compatible with those of his human keepers. No more rising at daybreak.

On the chilly day in February when Barney is eight months old, he is resting on the deck of his pool. Now grown to twenty-five pounds, or more than six times his weight at birth, he has changed from a soft, woolly, puppy-like beast to a miniature adult, sleekly furred and contoured. Aware of distant voices, he raises his head to see who's coming . . . only friends . . . Finn is talking to the volunteer and to the vet.

"We've tried a half-dozen kinds of marking devices. The tag I'm planning to fasten on Barney is stamped from anodized aluminum. It won't corrode in seawater. But Barney's not going to like it . . . are you, friend?"

The three enter the pen and, as the volunteer seizes Barney with gloved hands, Finn jabs the prong of a metal tag about the size of a fifty-cent piece, marked *17*, through the fleshy web of his foot.

"Males on the right foot; females on the left. When we're using binoculars in good light we can read the numbers from a distance of a hundred feet."

Barney shrieks in pure anger, and he shrieks again when moments later he is dumped into a carton of shredded paper. Carried to a truck, he is driven rapidly along the coastal highway to a fishing dock thirty miles north of Turtle Bay. Here two wardens are waiting on the bow of a patrol cruiser. At high noon, two hours after being "kidnapped," Barney is released among a dozen wild otters in a feeding raft offshore.

"We'll miss you, Barney," Finn calls, waving as the cruiser backs away, "but maybe we'll see you again. *Hasta la vista.*"

"Good-bye, Ten-Bucks-a-Day," says the veterinarian gruffly to conceal his real emotion. Yes, the cost of feeding Barney had risen sharply when he began to demand fresh seafood four times daily. The vet turns to the volunteer.

Her fine hair blowing across her face, she stares transfixed at the small brown otter growing smaller at the edge of the sea.

Meanwhile, Barney half-swims, half-crawls through slimy ropes of kelp in a frantic effort to follow the boat, then loses it behind a swell. He is terribly frightened, stunned by the vast light of the open sky and somewhat seasick, for he is unused to a bed that rises and falls, slides and halts. He calls . . . and calls again to surroundings empty of all things familiar . . . no echo returns.

Here the life of this little otter would have drawn to a close had not fate intervened. A female who was floating within earshot of Barney's piercing cries drew near and nuzzled him hopefully. A week earlier, in the half-light of sunset, her own pup had been struck by a boat propeller and instantly killed. When she grew weary of trying to support its torn and strangely unresponsive little form, she let it sink into the kelp forest. (No wild animal can know the meaning of death.)

So, when Barney cried for help she responded automatically, lifted him to her chest with her teeth, and began to lick his fur. No matter that he smelled unfamiliar; he satisfied an intense craving over which she had no control. Barney snuggled close, inhaling her breath and the faint tang of seaweed slime and shellfish juices that lingered in her warm wet fur. Through racial memory he knew that he was exactly where he ought to be . . . and wanted to be.

But for nearly a week he lost weight. He had to learn how to handle the rough seafoods that his new mother brought from the depths and deposited on her chest for him to share. When, however, he had grasped the trick of getting at the edible parts and discarding the shells, he began to enjoy the rich bounties of the sea. Some of them felt so *very* good as he crushed them between jaws in which permanent teeth were still erupting through swollen gums.

By his second week at sea, during spells when his mother was searching for food or was napping, he began to play with another large pup. The two wrestled and chased one another, rolled just under the surface, and pommeled each other with their fists. They interlocked mouths, doing no harm except to dislodge a milk tooth now and then. They keenly enjoyed their roughhouse play, unaware that they were perfecting the muscles and skills that later would serve them in boundary defense, in evading enemies, and in tearing abalones from the reefs.

To go back a bit, after Penny had left Barney in the care of the volunteer, she had walked along the shore for hours, trying to see into the future. More than once in her life had she turned to the open sea for guidance. Timeless, enduring, impersonal, it accepted her thoughts and returned the clearest among them.

Quite suddenly she had known her mind; she would study art.

So, on the following week she had entered the school of art at California Institute and, not long thereafter, realized that her choice had been wise. The mingled odors of turpentine, wheat paste, sizing, and wet clay that greeted her in the mornings when she entered the doors of the studios excited her in a way she had not thought possible. She studied techniques of sketching, painting, and sculpting; she experimented with the fire arts (ceramics and metal work); she took a course in anatomical drawing.

During the summer vacations she found work assisting the Turtle Bay veterinarian, whose practice had grown to include the care of wild animals. The medical problems that these presented were very like those of the farm animals and household pets with which he had been used to working daily. Among his new patients were deer crippled on the highways, hawks and gulls with wings broken by target shooters, starving seabirds besmeared with tarry oil, and seals entangled in scraps of fishnet. Penny helped to clean, dress, and dose these unfortunates, and she helped to release the ones that recovered, in a seashore park near Turtle Bay.

"Our percentage of failures is high," confessed the vet, "but that's not the point. Through *trying,* we show respect for life. I can't find the right words . . . but I think that anyone who cares deeply for animals must care deeply for people."

"No need to explain." Penny smiled.

So the years went by. Approaching time of graduation, Penny decided to build a career around her dual interests in wildlife and graphics. She would become a wildlife illustrator. Her chosen field would certainly be satisfying and it might become a paying profession. Having studied the artwork of others, she knew that illustrating animals is not easy. Although many artists can reproduce the profiles of an animal, few can bring the animal to life, complete with muscles and tendons that work, fur or feathers that drape as they should, and a posture faithful to some particular activity. In good illustrating there is an indescribable feeling of vitality.

It was her good fortune that when the editor of a publishing firm in San Francisco visited California Institute, scouting for talent, she had just completed a portfolio of drawings of Barney. Relying on photographs and her memory, she had sketched him begging for food with outstretched paws, sleeping like a fur piece carelessly thrown to the floor, sulking, splashing merrily in his pool, and in other ways filling the idle hours of puppyhood.

The editor was pleased with Penny's work. After consulting her teachers in the art school, he offered her a job as illustrator for a proposed new series of seashore books for young readers. Although the pay would be meager and the position probational for a year, Penny was delighted to accept.

Upon graduating, she leased a studio apartment in Monterey, where the sea winds could touch her face and remind her that she was at the place where land and water meet, at the place where lively images are certain to take form in the sensitive mind. Seldom now did she think of Barney. "Chances are he never made it," she mused one December evening as she stood at the window watching a distant play of lightning. "No one has seen an otter with tag number seventeen. But maybe he simply lost the tag. If he *were* alive he'd be . . . let's see . . . four years old, going on five."

January

On New Year's Day the ocean beside the breakwater at Turtle Bay is restless, disturbed by a storm beyond the horizon. At intervals a single swell rises above its fellows and crashes against the shore. The air is fresh, the winter sky is an envelope of clearest blue.

Penny is home for a visit and, at the moment, is preparing to scuba dive with her brother. She sits on the deck of her father's boat, the troller *Linda,* in the lee of the breakwater, sponging her mask. Her rubber suit is snugly in place; her swim fins are strapped. A waterproof sketch pad, a crayon, and a mesh bag for holding specimens hang from her waist. Her dark eyes shine. What new images, what new sources of inspiration for her artwork, will she find in the undersea garden that graces the rocks at the foot of the breakwater?

As she rolls backward from the edge of the *Linda,* the sun glare from wet seaweed strikes her eyes. She welcomes the soft yellow-green light below the kelp cover. She knows this kind of weed as the bull kelp. Each hollow stem is swollen near its upper end into a bulb like the butt of a cattleman's whip. Five fathoms down, the stem flares

17

abruptly into a holdfast, a sort of gnarled root, which cements the plant to its rocky base.

Penny swims underwater to where her brother is waiting, his body moving in harmony with the surge of the sea, a chain of bubbles rising from his mask. He has scouted out a place of singular beauty—a vertical wall of sandstone, rough-textured and carpeted with life. Upon its face a hundred kinds of plants and animals and a hundred others invisible to the naked eye play out their separate roles, each in the habitat best suited to its own needs for food, light, oxygen, and shelter. Some kinds are motionless while others creep, and still others dart from one feeding station to the next.

"It's all so balanced . . . so *right,*" breathes Penny to herself. Yet she can't shrug off a tiny touch of fear whenever she enters this dim, silent world. In its very mystery it seems a forbidden place. The waving sea vines are soothing at first, but in the end hypnotic.

The tapestry of the composition before her is a thin crust of limy algae dotted with pinky-purple bryozoans, brilliant orange sponges, and miniature seaweeds. Some of the weeds are fluorescent, casting rainbow hues from diffracting surfaces. Delicate stems of feather-boa kelps spring from crevices. Flowery white-and-green anemones wave their fingers, raking the currents for unwary prey. Red-and-ochre starfishes prowl at random, hunting for mussels, turban shells, or limpets.

A ballet dancer, thinks Penny as she pries a starfish from its bed and lets it spiral downward, arms outflung. She brushes jellyfish threads from her faceplate.

Now a thousand small silvery fishes appear and disappear. She knows them as tube-snouts, slim as pencils, shaped for quick flight from enemies.

Penny sketches rapidly. What a challenge she will face when she tries to recapture on a dry, two-dimensional canvas the pageant of life now passing before her!

As she approaches the garden to study its finer detail her eye is caught by a metallic gleam. It reflects from a metal pin, inscribed with meaningless numbers and letters, which someone has hammered into a crevice.

A marker, I suppose. Some biologist studying the movements of reef animals from season to season . . . or perhaps measuring the growth of seaweeds, she thinks to herself.

She drifts past the rock and into the shade of an overhang, where she finds herself staring into an ugly, rat-gray face. If the face of any wild creature on earth can be called ugly, it is that of the moray. Holding its undershot mouth agape, without moving its body, it returns Penny's stare. She fans her arms and retreats beyond range of its sharp teeth, for the mood of a moray can never be trusted.

Seeing that the gauge on her air tank shows three-quarters gone, she climbs to the surface, swallowing as she rises to ease the pressures in the air spaces of her ears.

Without warning she strikes a dark solid body, its outline blurred by her streaming mask. Until her vision clears she thinks it is her brother but, to her astonishment, she finds herself facing a sea otter. The animal periscopes for a long moment, rises in the water to its armpits, then dives and reappears to sniff at her face, thrust its nose under her chin, and gently seize one of her arms in its teeth. Incredible! She beckons wildly to her brother, who joins her. They remove their mouthpieces.

"Am I dreaming," Penny gasps, "or is it really Barney, alive and well after all these years?"

"Well, he's sure a tame one, but if he's Barney, how could he have known us? Could he have followed the scent of our bubbles? Do sea otters have that long a memory? Hey! Let's see if he's got a tag."

Cautiously the two grasp the animal's right foot and lift it from the water, exposing the web and the lower abdomen. He's a male, all right, but without a tag.

"Hold it! Here's a worn place in the fur between his toes . . . and two scars where prongs used to be. I'll bet a million he's Barney. *Hi ya,* Barney!"

In a quick fluid change the otter leaves their grasp, retreats several yards, and begins to smooth his fur with paws that blur as they move. (Penny thinks of a woman knitting.) Although he's willing to associate with these man-creatures, he doesn't care to be handled by them.

"I suppose we can never know for sure." Then a frown crosses Penny's brow. "If he stays near the breakwater he's apt to be run down by a boat. Anyway, let's come out in the morning and see if he's still here."

The divers return to the *Linda* while Barney—for indeed it is he—resumes his search for crabs in the dark crannies of the sea.

To go back in time, when Barney was released on that February day four years ago he was at the point of being weaned. As spring and summer came he saw less of his adoptive mother and more of his age-mates. In his second year he joined a group of otters who were roving in desultory fashion southeastward along the coast in advance of established breeding colonies. These pioneers were re-populating the food-rich waters from which otters had been exterminated by man's guns, spears, and nets nearly a century earlier. Like the human pioneers of the American West, most were bachelors; a few were accompanied by females. Barney's group—collectively known as the "front"—would forage back and forth within a radius of three or four miles until they had eaten all the easily gathered sea urchins, then would move on to new pastures. They often stayed four or five months along one section of the coast before leaving it.

They favored urchin foods simply because the local species—the giant red urchin and the purple urchin—were both tasty and easily captured. Barney learned to knock them from the rocks or from the kelps where they were grazing. Then, when he had risen to the surface with two or three tucked under his forelegs, he would bite quickly through the thin shell of each, slurp the brown-and-orange contents, and dive for others. In the wake of the front group the sea bottom was strewn with urchin fragments, like egg-shells after a kindergarten picnic. On some of these shards the spines continued to move feebly after death, obeying the last order dictated by a primitive mind.

Life beyond the breakers was not altogether free of tragedies. Barney's first playmate, the male with whom he had sported while both were pups, disappeared in a gale. The unlucky youngster was blown ten miles offshore at night on a mat of seaweed. At daybreak he saw the looming of the land and tried bravely to rejoin his companions but, before he could reach them, swimming upwind, he grew chilled and exhausted. He died face down in the sea.

Sea otters do not, as do seals and whales and other marine mammals, store energy in a blubber layer beneath the skin, nor do they cache food against time of need. Their life is a daily, never-ending struggle to stoke the internal fires which ward off the penetrating chill of the sea.

During his first four years, Barney's activities revolved around feeding, resting, grooming, and playing. On rare occasions he would leave the ocean to haul out on near-shore rocks or on some secluded beach where, between tides, he would sprawl for a while in the sun.

As he grew older he grew in understanding of the social structure of his group. When, time and again, he had risen to the surface with an abalone or other prize and had seen it snatched from his chest by an older animal, he learned to feed at a safe distance from the dominant males. Such thievery was wholly impersonal; Barney himself practiced it on younger animals. It *was* annoying, though, to lose one's lunch to an otter who was already carrying food.

Barney usually slept at night in a community raft from which he paddled away soon after daybreak to commence feeding. He fed rather steadily until early afternoon, when he rejoined his companions to groom, sleep, and play. As he grew adept at finding food he began also to feed at night. How he was able to do so is not clear. Certainly he used his sensitive paws and whiskers to probe for food; perhaps he relied also on the weak light of the stars and the moon, as well as the ever-present cold, greenish shine of bioluminescent organisms. (Susie, the first sea otter to survive in a zoo for more than a few days, was totally blind during her last year of life yet remained in good health. She located by feel the foods that were tossed into her pool and, moreover, readily identified her favorites among them.)

Barney seldom ventured beyond the kelps or into waters over ten fathoms deep. Whenever he approached the normal limit of his feeding range he grew uneasy, fear-ful of dangers unknown. And the deeper the water, the more time-consuming and tiring became his dives. He had learned genetically to know his place, the place where his chance for survival was greatest.

In the summer of his fifth year he grew restless. (He could not know that his testes were swelling with ripe sperms and were secreting androgens, or chemical messengers, into his blood.) He was on the threshold of sexual maturity. Certain tantalizing odors reaching him by way of the north-westerly winds heightened his irritability. His play with other young-adult males often ended in passionate wrestling matches in which each tried to bite the other's face or to ride his body.

So, in early winter Barney turned northwestward to-ward the waters of his birth. He had memorized the salient landmarks and seamarks along the route—the looming headlands with their flashing beacons, the submarine reefs and canyons, and the sweet currents of water fanning out from estuaries. He swam at times on his back and at times on his belly like a seal, boosted by the steady current that sets toward Oregon during the winter. Traveling, feeding, and resting, he covered a hundred miles in two weeks.

On the first of January, rising high on the crest of a swell, he sighted the breakwater at Turtle Bay and knew that he was home. But now he found himself among a score of otters, while at the time of his birth there had been only himself and his mother. Moreover, many of those present were females with pups, and a few were females in heat— or so his nose told him.

All at once he picked up a faint, nostalgic scent that turned his head toward the breakwater. The gabbling of gulls above his head seemed suddenly the voices of man-creatures. His pace increased as he neared the breakwater and the black-garbed "mother" who had cared for him during the most sensitive weeks of his life. He pushed into Penny's arms.

Early on the second of January the *Linda* leaves her mooring in Turtle Bay and once again chugs toward the open sea, carrying Penny, her brother, and Finn Peterson. Over coffee and crullers the three make plans.

"Supposing the otter isn't Barney—could it be a sort of simpleminded critter—one that doesn't know any better than to approach people?"

"I doubt it. Nature has no room for fools. The struggle for existence among wild animals is so keen that, when a fool is born, he or she doesn't survive long."

"Well, I think the friendly otter *is* Barney. I'm going to feed him some squid from the bait box," says the brother.

Finn hesitates. "Couldn't that lead to trouble? You saw the TV movies made up north last year. The photographers were on location for a week before they started shooting, and they trained two wild otters to take food from a diver's hand. They got spectacular underwater footage, all right, but a month later we found two dying otters hauled out on the beach. Both had wounds that might have been made by a spear. Although I have no proof, I suspect that the animals were done in because they had learned to trust people.

"I've been thinking about the possibilities for research," continues Finn with half-closed eyes. "By getting close to a tame otter we could see what kinds of shellfish, and in what amounts, he was feeding on. We could clock the timing of his dives, of his grooming periods, and so on. We might even be able to wire him with a miniature radio transmitter that would let us follow his movements night and day."

"Not on Barney, please," counters Penny. "He's my only child. Try your radio on a wild one. I've no objection to your *studying* Barney's behavior as long as you don't *instrument* him. And while you're studying, I'll be sketching."

The *Linda* begins to rock gently as she clears the breakwater and feels the toe of the sea. Finn, scanning the kelp from the wheelhouse, points to a solitary otter swimming on its belly toward the boat. He shouts, "A hundred yards off the port bow!"

The animal halts, lifts its head and, like an Indian on the western plains, raises one paw to shield its eyes. It continues on course, then comes to a gliding stop within twenty feet of the watchers. Finn cuts the engine. The otter begins to groom itself, glancing warily now and then at the boat. Finn studies it.

"He *could* be Barney. . . . He's the right size for a youngish adult male . . . about four feet long from snout to tip of tail . . . forty to fifty pounds. He's getting a lot of white hairs around his face, but he's not yet a white-headed adult."

Barney demonstrates a sea otter's unique skills in contortion. For eight minutes, by Finn's watch, he somersaults in a tight circle, repeatedly scrubbing every square inch of his body. He pauses at intervals to blow into the fur of his chest or to lick his paws. For two minutes he lies on his back and "logrolls" from side to side. He squeezes water from his fur with his paws. For six minutes he works on his hind feet and tail, bringing his heels, incredibly, to the back of his neck. Turning himself into a ball, he cleans the base of his tail with his teeth. At each turn of his neck the fur breaks into silver arrows like shafts of a waterfall. In a four-minute act he adds a pat here and there and massages his face and jaws. At last he drapes a ribbon of kelp across his chest to serve as a stabilizer, rolls belly-up, with head and feet in the air, and instantly falls asleep. Sunlight striking his wet groin throws sharply into relief his maleness—the bulges of testes and the stout, bone-stiffened penis under the skin. He floats in a transparent hammock.

"What a performance," whispers Finn, stowing his camera. "I shot two rolls of film."

The tide carries the *Linda* slowly away from Barney. Penny's brother stands watch at the wheel while the other two adventurers squeeze into the galley.

Penny's quick mind has darted ahead. "If Barney's actions this morning were typical, sea otters must burn an awful lot of energy every day. They must eat far more than the average wild animal."

Finn responds. "That's hardly the kind of problem that the agency is prepared to study. We're a practical outfit." He pauses. "Still . . . we're getting complaints every day from clammers and lobstermen. They resent the otters coming back to inshore waters where humans have long claimed exclusive right to harvest shellfish. Maybe I *can* interest the agency in funding a study of otter food requirements."

His thought bears fruit. Arrangements are made to capture two healthy otters in Monterey Bay for physiological study. In the slanting sunlight of a January evening, Finn and a crew of helpers spread a gill net three hundred feet long in waters where otters are known to move between feeding and rafting sites. The helpers are not optimistic, for none has ever tried to catch an otter in the open sea—in its own element.

Approaching the net at first light of the next morning they see dark shapes along the float line. "We've done it!" exults Finn, but as he reaches the first otter he finds it cold and still, barely awash in coils of webbing. When he lifts the body he finds beneath it the quiet form of a six-foot blue shark. Thoughtfully, he reconstructs the tragedy.

"The shark must have hit the struggling otter early in the evening and soon found *itself* tangled. It kept trying to dive until both animals suffocated. A shark has to have active circulation of water across its gills."

But three other shapes along the line are otters very much alive. They hiss and screech and fight the net, enmeshing themselves so thoroughly that the crew must cut them loose strand by strand.

"There's got to be a better way of doing this," Finn frowns. "Anyway, let's keep the old male and the female and let the young one go. He's home free."

The crew place the two captives in individual crates and truck them to the Marine Station, where arrangements to receive them have been made. For several weeks each otter will be fed weighed amounts of food, while at intervals its metabolic rate—its rate of internal heat production—will be measured in a closed chamber. The payoff figure will be the amount of oxygen used by each animal per minute per unit of body weight.

The Turtle Bay veterinarian has offered to spend a day at the station to insure that the otters are properly housed. Finn visits regularly to supply clams, squids, crabs, and fish fillets for their meals. On the last day of January the physiologists review their findings.

"The sea otter is a peculiar animal," explains one of them at a noonday seminar. "Its metabolic rate is three to four times higher than that of a land mammal of the same size. Its energy need is correspondingly high. Our test animals ate every day at least *one-fifth* of their own weight. A *wild* otter obliged to work for a living would eat even more—perhaps one-third of its weight. It's safe to say that an adult otter consumes five to six thousand pounds of food a year. Putting it another way, a fifty-pound animal needs the food equivalent of about five thousand calories a day, or nearly three times the need of an active, growing, fifty-pound boy or girl.

"The harbor seal, a marine mammal that occupies an ecologic niche somewhat like that of the sea otter, eats only one-twentieth of its body weight per day. The main reason why the seal's life is metabolically less demanding must be that the seal's body is insulated by a thick coat of blubber. Whereas the sea otter constantly has to spend energy to maintain the insulating quality of its fur, the seal has only to eat and grow fat."

His eyes twinkle. "Peterson and I have roughly calculated, for the benefit of those few Californians who hate sea otters, that a fifty-pound individual could polish off every day twelve small abalones, twenty sea urchins, eleven rock crabs, sixty kelp crabs, and 112 snails, or fifteen pounds of seafood in total.

"And, by the way, in dissecting the bodies of otters found on the beach we had already learned that the liver is surprisingly large—three or four pounds. Body weight for weight, it's at least twice as large as the liver of other marine mammals. Since the liver is the main organ for producing and releasing the breakdown products of food, its large size is quite certainly tied in with the otter's rapid metabolism.

"To return to our study of the two live animals, we learned, as we had suspected, that an otter's digestion is very rapid. We hid a spoonful of red food-color in clams and fed the clams to the captives; the red showed up in their droppings only three hours later."

The speaker pauses. "While most of us agree that the sea otter is a charming creature, we should also admit that it is hyperactive and greedy, and that its dining habits are sloppy. It can't touch man for being wasteful, though, for every scrap that the otter drops into the sea is recycled."

He concludes. "Perhaps in the otter's high food demand we have a partial explanation for the very slow recovery of the California population since 1913, when it was first given legal protection by the state. It has grown at an annual rate of only 5 percent during a period of nearly seventy years—and lately the rate is slowing down. The slow recovery is telling us, I think, that the life of the average otter is strictly ordered. The otter leads a knife-edge existence."

February

There in the kelp beds were the Sea Otters carefully
wrapped up as though they were in bed, all lying
on their backs in the water with their heads sticking
up and their paws together. They looked like a con-
vention of bishops in mud baths in Baden Baden.
They were quite the most enchanting animals I
think I have ever seen.

—GERALD DURRELL

Precisely so had Barney settled his sleek body when last he
was seen by the *Linda*'s passengers near the breakwater.
Now he has resumed his search for a female in heat among
the otters rafted off Turtle Bay. He has learned that all rafts
are not alike in age and sex composition. Some contain
mostly young males from two to six years old, while others,
especially those in sheltered waters, contain mature females,
pups, and a scattering of full-grown territorial males.

Their territoriality is weakly expressed. How could it
be otherwise in the shifting, three-dimensional sea? Evidence of territoriality is supported by the fact that males
outweigh females by about 1.3 to 1, a characteristic of
animals in which male dominance is unmistakable.

31

Barney circles a raft from which an alluring scent lingers in the air, then swims cautiously upwind to within five yards of a mated pair. The female sees him first and rises in alarm; the male growls; all three dive. When they surface, Barney and the male are tumbling as one, spattering the sea with molten silver. For a minute their bodies blur. Then the antagonists separate, rise high in the water, and spar like boxers. Barney retreats a respectful distance, breathing noisily and shaking his head. At the start of this adventure, fighting had been far from his intention. He tries another strategy, shooting belly-down toward the pair, but is forced to withdraw with a bloody nose. As though thinking, *It's not worth it,* he swims slowly to a kelp bed near the breakwater. Life has betrayed him.

After grooming his pelage, he begins to dive for food in the sun-dappled shallows, for he is ravenous. Lifting his head high, he hurls his body forward and downward in a single turn, his webbed feet flashing against the sky. Bubbles mark his descent.

On the bottom he threads his way among kelp stems, now crawling for a few paces, now swimming. A muted chorus from surf and shore-break drums against his ears. Over sandy patches he paws actively for clams, stirring up motes that cloud the surrounding water. When a clam retreats too fast for his paws to follow he bites off its tasty "neck" or siphon. Now he brushes through a colony of pale sea pens writing in the sand. Alarmed by his shadow, a feather-duster worm retracts its red-wine gills. Along rocky ledges he probes for abalones, mussels, scallops, rock crabs, jingles, limpets, starfishes, and chitons.

He makes a lucky find—a red abalone ten inches long. He tries to pull it free, but it presses its powerful suction foot against solid rock. Having been below for two minutes he rises for a quick breath before diving again. He tugs and pries vainly at the mollusk until three minutes have passed—the comfort limit of his breath-holding ability. On a third dive he seizes a heavy granite rock the size of a man's shoe and begins to pound the abalone with it. *Chunk! Chunk! Chunk! Chunk!* . . . a pause to tug at the shell . . . *Chunk! Chunk!* . . . fragments fall from the edge. After three minutes' hard work he succeeds in jarring the creature from its base.

He rises, puffing, with the prize under his right foreleg, turns belly-up, and places it upon his chest. As he bites off chunks of white meat he rolls from side to side, splashing water over his body to clean both the food and his pelage.

He throws inedible fragments, as if they were peanut shells, right and left into the sea.

On his next dive he digs three heavy clams, then selects a flat two-pound rock and carries the lot to the surface in the peculiar pouch or fold of skin that crosses his chest and serves as a pantry. Settled comfortably, he places the rock on his chest and, clutching a clam between both paws, raises it high above his head and brings it down upon this rocky anvil in quick, repeated blows. He had started to practice this trick when he was a pup, drumming his little fists on his chest in imitation of his mother's actions. The noise of his pounding spreads widely and attracts a gull, who has learned to pester feeding otters in the hope of snatching crumbs. Barney makes a quick swipe with his paw and the bird, minus a tail feather, retreats in alarm.

Down and up, down and up through the swaying stems that Shelley called "the oozy woods which wear the sapless foliage of the ocean," Barney feeds for an hour. He tops off his meal with a small octopus he finds lurking in an empty can. He had seen hundreds of aluminum cans on the ocean floor before he learned, by watching older otters feed, that often a can contained an octopus, which he could release by crunching the thin metal with his teeth.

When Barney's remote ancestors hit upon the idea of using a rock either as anvil or hammer they became the first nonprimate tool-using mammals on the planet. Modern sea otters retain that distinction. The trick is hardly more "intelligent," however, than is the maneuver of the raven, which drops a clam from the air to break it upon the rocks, or of the Galapagos finch, which carries a thorn in its beak to pick a woodboring grub from its tunnel, or of the captive blue jay, which makes a tool from torn paper to reach a tidbit too distant for its beak, or of the African vulture, which breaks an ostrich egg by bombarding it with a pebble hurled from its mouth, or of the mongoose, which breaks a millipede by throwing it between its legs against a rock.

Arville P. Whitt, a college student, once watched a young male sea otter in an aquarium pool playing for an hour with several rocks. The otter repeatedly tried to break one rock against another one that was resting on his chest; he struck the two rocks together as one might strike cymbals, pounded the concrete edge of his pool with one, and bit one with his teeth. "Unable to break any of the rocks," the student wrote, "he finally went to sleep with two tucked under his left front limb." An intelligent performance?

An otter feeding at sea will often carry a favorite rock under its arm through successive dives, a habit as curious as its use of the rock in the first place, for the act hints at anticipated use. It suggests that otters may have that special sense that nonhuman mammals are not supposed to have— a sense of the future. It would be foolish, though, to believe that otters will ever advance to the evolutionary stage of making and storing tools as do humans. On the contrary, the otters have dug themselves into an evolutionary groove out of which they will doubtless never climb—or climb very far.

Biologists have asked, "Why did tool-using by non-primates evolve *only* in sea otters?" The answer lies in the singular ecologic niche that Barney and his companions occupy. They depend for food on hard-shelled organisms,

yet float on a fluid medium—water—that offers no hard
surfaces against which the organisms might be crushed or
opened. What is more logical, then, that the ancestral otters
should have learned, early in geologic time, to pound two
shellfishes together and, later, to pound a shellfish against
a rock?

But why don't *seals* use rocks as tools? They, too—or
at least some inshore species—feed partly on shellfish. The
answer is simply that ancestral seals and ancestral sea otters
took different evolutionary routes. It can never be known
why their routes were unalike, only that they must have been.
Through competitive natural selection the bodies of the
early seals became slender and streamlined, adapted mainly
to hunting fishes, squids, and swimming crabs. Through
the same process the bodies of the early sea otters became
compact and buoyant, adapted to foraging on the bottom
and to floating for long periods between dives. The fact that
both the modern seal and the modern sea otter, feeding side
by side in the same kelp groves, often hunt the same species
is beside the point. One might say fancifully that the ancestral seals never "intended" to exploit the ecologic niche
that was later to be exploited by the sea otters. The ancestral
seals never "intended" to feed on abalones and heavy-shelled
clams and thus never acquired the body architecture that
would have enabled them to do so.

Whereas a dog or a cat has front feet that are not greatly
dissimilar in shape and function to its hind feet, a sea otter
has front feet that are strikingly unlike its hind ones. In
terms of their functions, the front feet are hands, while the
hind feet are paddles or hydrofoils.

"Their fore feet," wrote army surgeon Elliott Coues in

1877, "are remarkably small, giving the limbs an appearance which suggests amputation at the wrist."

What about this sea otter hand? Is it the kind of dextrous, multipurpose structure one associates with a tool user? Not really. It is basically the paw of a land carnivore. It gradually changed to meet the peculiar needs of the sea otter by developing a thick, pebbly-rough palm—the better to capture food—and skin creases backed by appropriate muscles and tendons—the better to hold food. It became a sort of hand within a mitten. All things considered, the sea otter hand, although lacking man's divided fingers and opposable thumb, is an extremely sensitive and useful member.

Ron and Joyce Church, pioneer undersea photographers, once hid a number of sea urchins on a rocky bottom, then waited nearby in scuba suits to catch an otter in the act of feeding. When an obliging subject appeared

> . . . he started feeling along the rocks. He was actually unaware that any urchins were there until he felt one with his paws or whiskers. . . . He repeatedly shoved his forepaws into each crack and crevice, touching and feeling everything.

Far to the northward on this February day, **Penny** is seated on the steps of the Marine Station sharing a bag lunch with Amelia Caring, the librarian of the station. Penny's visit has a purpose.

"I've been talking with friends in Otters Alive. We'd like to harness the energy and enthusiasm of the people who call at the Otters' office wanting to get involved in research or asking how they can help promote otter conservation. We'd like to mobilize their individual skills and talents. Do you suppose they could be put to work as amateur biologists? Could they watch otters through binoculars and make life-history records? I've read everything you've given me about sea otters and it seems there's much yet to be learned about their biology. For example, no one has described—and possibly no one has ever seen—the actual birth of an otter in the wild."

"It's odd that you should ask," Amelia purses her lips. "I've been thinking along the same line. The amateurs could make useful observations and they could comb the beaches for stranded animals, dead or alive. If they found a sick one they could take turns being 'otter sitters' until a vet could be called. I'd be willing to act as a clearinghouse for information, receiving otter news and passing it along to the Fishery Agency or to my colleagues here at the station."

"Great!" says Penny. "We would want to include the Turtle Bay veterinarian in our program. He knows a lot about the care of stranded animals."

She continues. "Wouldn't it be better if the clearinghouse were at the headquarters of Otters Alive? You're only one person, while the Otters have people in the office all day, and the secretary lives in a nearby apartment where she can be reached in an emergency."

"I suppose you're right."

The two toss ideas back and forth as the shadows of the cypress trees move slowly across the grass. A squirrel, its nose to the ground, searches for food at their feet. A week later they meet with representatives of Otters Alive, the Fishery Agency, the California Divers Club, and two scientists of the Marine Station, including a seaweed expert.

The product of the meeting is The Network. Composed

mainly of students and young persons, it also includes a number of oldsters—retirees living near the coast who regularly visit the beach to collect driftwood and other gifts of the sea—or simply to walk in the sun. Through long acquaintance with the shore they have learned to love it for its wild beauty and to respect it for its power to move them. Proud of California's sea otter colony, they are eager to contribute what they can to knowledge of otter biology.

The Network's first contribution comes from a strange quarter. The caretaker at the Marine Station remarks to his daughter that he has often seen a solitary otter resting at night on a secluded beach near the station's campus. The girl relays the information to her biology teacher and the teacher, knowing that otters seldom haul out on land, drives to the station that very evening. He walks quietly to a point near the beach and hides in a myrtle thicket.

The incoming tide, illuminated by a full moon and the fading light of day, creeps toward him. Riding its flow is an otter that, from its large size, can only be a male. Now the animal swims parallel to the shore, fixing his eyes cautiously on the land. He emerges from the water and humps awkwardly toward the upper sands. He walks by arching his back and pulling his body forward with his paws, alternately placing his webbed hind feet lightly on the sand and letting them drag. He settles on one side and begins to nuzzle his fur.

The teacher finally stretches his cramped legs and slips away. He returns at dawn before school to examine the haulout beach. Here is the imprint of a hind foot, a depression shaped like an oval fan with five nearly parallel marks. These, he thinks, would be the five toes. Here is a forepaw track as round and simple as one made by the tip of a cane. Here is a sinuous line traced by the tip of a dragging tail. And here is a fresh dropping about four inches long, shaped like a wiener sausage, crammed with coarse shell fragments. Other droppings, bleached and crumbling, lie above the tideline.

"Chance favors the prepared mind." He recalls the words of Pasteur. Poking a dropping with a twig, he thinks, *Here's a unique opportunity to learn what sea otters eat.*

As a consequence of his idea, members of his biology class take on the task of visiting the beach every morning to collect fresh droppings in plastic bags. They learn to distinguish the true ones from the lookalike pellets regurgitated by gulls. They label the true ones and freeze them for study. Although their classmates tease them as the Dawn Raiders, and though they joke among themselves at their odd assignment, they understand its biological value.

Weeks later, two hundred droppings are analyzed by specialists at the Marine Station. From each sample—after it has been washed and screened—fragments of shells, spines, bristles, squid beaks, and fish bones are carefully removed and identified by species. This detective work is not as hard as it might seem, for the diet of a sea otter is monotonous. One who examines an otter's droppings soon learns to recognize the food species they represent.

The students learn that the diet of this particular animal included rock crabs, spider crabs, sand crabs, leaf barnacles, isopods (or sea slaters), mussels, squids, turban shells, scallops, tube worms, starfishes, sea urchins, a few unidentifiable fishes, and one distinctive, flesh-colored worm known as the fat innkeeper.

"Of course," the teacher explains, "any animal's diet is influenced by the foods that are easiest to gather. From the fact that our samples contained few sea urchins we can deduce that urchins are scarce in the bay. And the reason they're scarce is probably that the otter herd that moved in here several years ago reduced their numbers."

To wind up the project, the teacher reports the findings in a journal of zoology, giving credit by name to all the students who helped.

The seaweed groves or hanging forests through which Barney roams day and night are organic systems well-known to only a tiny fraction of humankind, yet they flourish in all the cool marine waters of the world. Some of the longest kelps stretch two hundred feet from holdfast to tip of longest frond. The giant kelp grows faster than any other species in the plant kingdom, including the fast-growing tropical bamboos. A stem may spurt two feet in a day. Under favorable circumstances the productivity of a kelp grove—about twenty-five tons per acre per year—is as great as that of a lush tropical forest or a sugarcane field under intensive cultivation. Over half the kelp biomass produced by a typical grove is exported as detritus, and thus the grove supports life far outside its boundaries.

The complexity of life in a kelp grove is greater than that in most land forests. Two biologists who studied a grove smaller than an acre identified eighty species of algae, three hundred species of large invertebrates, and sixty species of fish. Every kelp plant has a second species growing upon it, and most have a third growing piggyback upon the second. All of the great groups, or phyla, of marine animals are represented in the groves. Some species feed directly on the kelp tissues, some graze on the oozy litter of the sea floor, some use the weak shifting shadows of the groves as pirate lairs from which to launch predatory raids, and some use the groves only seasonally when they come in from offshore waters to spawn. Here these animals began to coevolve long, long ago and here they now survive.

"No human ear heard the first sea lion roar," Don Graeme Kelly, California artist, has written, "or heard the first gull cry on the shore."

During the evolution of the kelps in hundreds of millions of years, only two mammals are known to have had a marked impact upon them. One was the extinct Steller sea cow, a lumbering beast related to the Florida manatee, though much larger (weighing up to seven tons). It wallowed along the edges of the North Pacific Ocean, feeding wholly on kelps and surfgrasses. Prehistoric man is blamed for wiping out all but a relict few hundreds. When these few were discovered in 1741 they were living on two unpopulated islands in the Bering Sea. Twenty-seven years later, hunters had killed the last one.

The impact of sea otters upon the kelp beds, although less direct, is more impressive than was the impact of the sea cows. The otters, by removing the sea urchins and snails, which feed upon kelp, allow the kelp beds first to recover and later to flourish. Along certain Aleutian Islands where sea otters have been abundant for decades, brown kelp fronds cover the surface of the sea in continuous carpets. The sea otter is rightly called a keystone species in the structure of the inshore marine communities where it lives.

March

Penny's father and brother have piloted the troller *Linda* to Monterey to be fitted with a new compressor. On a Sunday morning the two are loafing on the gray planks of the city dock, talking to a fisherman who is mending a crab pot. Penny has joined her family for the day, and Amelia Caring is with her.

"The clamming at Arenosa Beach is shot to hell," mutters the pot mender, jerking at a length of wire. "The damn otters moved in there two years ago February and now the beach is covered with empty shells."

But Amelia has studied the evidence pro and con as to whether harm has been done by otters to California's shellfisheries. She tactfully suggests, "There's no doubt they do eat clams, but I'm wondering if the real problem isn't too many people. I was clamming at Arenosa on the Fourth of July a year before the otters got there, and afterward I read that over a hundred thousand people were on the beach that same weekend. Most were clamming."

"When I was a kid," puts in Penny's father, "the bag limit was a hundred clams; now it's only ten—and a man's lucky to get three or four on a tide. Abalones, too, are

getting scarce, but I must admit they're scarce outside, as well as inside, otter country. The Crabbers Union says that the otters are eating three million dollars worth of shellfish a year."

Amelia scribbles on the margin of a newspaper. "Given fifteen hundred otters, young and old . . . annual consumption per otter three thousand pounds . . . value of shellfish in the wild at, say fifty cents a pound. It comes to 2.2 million dollars. Of course, any estimate will depend on the value you assign to shellfish in the ocean, uncaught. And the diet of the otter is very broad—more than fifty kinds of fish and shellfish. How are you going to put a price on the lot?"

A warden in uniform saunters up and enters the conversation. "I couldn't help listening. Speaking for the agency, we'd like to stabilize the otter population at two thousand animals—which would satisfy the commercial fishermen and yet leave plenty of animals for the otter lovers. And we'd like to restrict the otters geographically to one or two refuges."

"What do you mean, 'stabilize' the population? Do you plan to kill off the surplus every year?"

The warden circles the question. "We're keeping our options open. We've started a program of live-capturing and moving animals away from important shellfish grounds. It's not too successful yet; about a third of the first lot came back. We've sent a few animals to zoos and aquariums.

"And we're approaching the problem from another angle. On the chance that we *won't* be able to contain the population—at acceptable cost, anyway—we're starting to raise abalones in tanks. Aquaculture. The Japanese have been doing it for years."

"In other words," says Amelia, "you're going to leave the wild abalones for the wild otters."

"More or less. Let's face it. Abalones are a luxury food. We can't justify killing otters simply to encourage the growth of abalones. Too many people love otters."

The pot mender lights a cigarette and speaks through a cloud of smoke. "I was an abalone diver in '28 when there was only eleven of us had commercial licenses. Later on, there was over *eight hundred.* Now the agency holds the number to two hundred. Anyway, in my opinion there's three reasons why the abalones are scarce: too many divers, the bag limit's too high, and too many damn otters."

"It seems to me there's an argument you've all overlooked," offers Amelia. "It's on the positive side. The otters, by eating sea urchins, which feed on kelp, help to maintain the kelp beds. To a degree, the otters control their own environment. You'll admit that, without its kelp beds, California would have fewer fish."

"The abalone divers made a suggestion that we listened to but rejected as screwy," adds the warden. "Their idea was to *encourage* sea otters along a stretch of the coast until the animals had cleaned out the urchins but not yet the abalones. At that point in time the agency would relocate the otters. In the absence of urchins, the kelps would recover, followed by abalones, followed by divers."

Amelia chuckles. "That's similar to the way country squires in England mow their lawns—move a flock of sheep from one to another. I can see why you rejected the idea. Moving sheep is one thing; moving wild otters *en masse* is another."

Penny's father turns to Amelia. "You're right about the kelp groves being good for fish. We see schools of anchovy there, and bonitos chasing the anchovy. . . . Lots of kelp bass and sheephead . . . rockfish, perch, and croaker . . . sometimes barracudas . . . and millions of little buggers that I don't know the names of."

The fisherman intervenes. "I've heard over and over how the otters make the kelp grow. Maybe they do, but down at Valencia Reef, where there never was any otters, the kelp started to grow for no reason at all and in three years it was so thick you could hardly stick an oar in it."

"In the ocean," says Amelia, "there aren't any simple causes. Maybe the seawater temperature at Valencia Reef changed. Maybe the City of San Diego changed the treatment of sewage it was dumping into the coastal waters there. Who can say?"

Penny has been listening without comment. Now she breaks in.

"What would happen if the otters were simply left alone? Would they suffer? Would they reach a population ceiling and then starve?"

Amelia replies. "Not really. Their numbers would be controlled by nature, as always. When the population approached a certain upper limit it would begin to press harder and harder upon its resources. The average otter would have to hunt longer and longer every day for food, at greater cost in energy. Its resistance to the parasites that are always present in its body would weaken. Its body size would decline and its fat reserves would disappear. As a consequence, the average young animal would reach puberty a year or two later than it used to, and would be less likely to survive winter storms. The incidence of stillbirths would rise. Population growth would grind slowly to a halt."

Penny rises to her feet. "I hate to think that Barney will ever suffer, but of course he's bound to before he dies. Hardship, or environmental pressure, is the thing that shapes the otter species—just as the pressure of the artist's fingers shapes the wet clay."

"Look, Penny, we're not in the Garden of Eden." Her father puts a kindly hand on her shoulder. "We're in the real world. What will happen to abalones and otters will be decided by the politicians in Sacramento, not by Mother Nature."

"Okay, Pop, but you can bet that we otter-watchers will be watching the politicians. If they get out of line, they'll hear from us."

The pot mender pauses and stretches his arms above his head. "I still don't like the damn otters. It's hard to like animals that take food from my table."

Deep in the sea ten fathoms below the cold shadows of the Santa Lucia Cliffs, Barney and a female companion are sliding through the ruins of a wreck. Here the bones of the *Esperanza* have lain for a century, marked on a yellowing chart by a cross and a question mark (?), and later forgotten. She was a coastal schooner carrying hides, and she died unseen on a black winter's night at a place where basalt columns rise from the water like broken teeth. No human diver has found her yet, for none has dared to venture among the channels where the green water surges endlessly back and forth between the cliffs and the ragged rocks.

Barney's companion is a six-year-old who has not yet come in heat and never will, for her ovaries were destined through accident of birth to remain infantile. She and Barney enjoy each other's company; neither is a threat to the other. They are, if one may put it this way, temporary friends. As a side effect of the female's sexual imbalance, her features are strangely like those of an adult male. She is nearly white from head to chest, gray on the belly, and tinged with gray along the underside of the tail.

The two companions expect to find a tasty octopus lurking deep in the skeleton of the ship. In descending they had pushed through a rubbery jungle of sea palms, those stout, stubby, treelike algae that in turbulent waters replace the more fragile kelps. The otters had skillfully bypassed a tangle of monofilament fishnet, weighted with leads, which had been drifting in the deep currents until it hung up on the timbers of the wreck.

Now as they enter the gloom of the dead ship their eyes dilate. They swim more cautiously, although they feel no fear, for their everyday world is one of shifting colors, of haloes and spectres, and of looming and fading images.

The timbers of the *Esperanza* have the outline, but not the substance, of oaken beams. Long ago they were riddled by teredos and gribbles and, as they decayed, they became overgrown by crusting algae, corals, and limy tube worms which in community formed an exoskeleton. Of the ship's log, the cargo of hides, the barrels, the rigging, the men, and the ship's cat no particle remains; all are now molecules of the world ocean. The fittings of the masts and capstans are gargoyles of rust. Along the keel lies a pile of greenish boulders uniformly the size of melons—the ballast rocks. Here they will lie for centuries after the *Esperanza*'s silhouette has disappeared.

As Barney's whiskers probe the undergrowth they startle into flight tiny crawlers, which leap into watery space behind him. At one point he presses through a cloud of pale, nameless swimmers that flutter in total silence as they give way for his body.

He stops . . .

Barely visible in the dark is a great nose like a swollen bag—a warty nose splotched with purplish brown and gray, surmounted by heavy-lidded eyes with slit pupils. Extending from the "nose"—which is really a body—eight long arms studded with rows of white suckers drape themselves in curves along the walls of the creature's lair. It is an octopus the size of a small washtub. Here it sits, anticipating the night, when it will glide silently over the sea floor in search of fish, shellfish, and carrion.

Barney hangs weightless for half a minute, staring at the thing while crude reflexive thoughts race through his brain. Although he has never seen an octopus this big, he knows from painful experience that even smaller ones can inflict a painful bite, followed by burning pain.

The octopus shoots a burst of ink, which is instantly a brown cloud. Barney comes to a prudent decision. He makes a U-turn and thrusts his webbed foot against the water. The foot flares into a paddle and shoots him upward to the light.

After napping side by side on an undulating couch of weed, Barney and the female move along the track of the sun to a sheltered cove, where again they begin to feed. Both are aware that man-creatures are on the beach. The knowledge does not trouble Barney, although it makes the female nervous. She lifts her head often to scrutinize the strangers. The two animals work their way casually into shoal water and, as they do, a scuba diver dressed in black-and-orange swims underwater toward them, following a line of concrete blocks that he had anchored on the bottom a year earlier. These are markers along a transect line. He is a biologist monitoring the repopulation of sea urchins to a measured plot from which all urchins had been removed. He nears the end of the transect.

Whoosh! A dark shape breaks the light above him and icewater pumps through his veins. A shark? A giant ray? A killer whale? None of these—only Barney in a playful mood.

"Jeez! I had a scare out there by the last marker," he later tells his buddy on the beach. "I thought my number was up. This thing came right up to me . . . a big otter. He poked me with his nose and felt me with his paws and started to nibble on my regulator. I wasn't sure of his intentions. Anyway, I climbed to the top and he seemed to lose interest."

"That's one for the record," says his diving buddy. "Well, we're finished here for today. I doubt if the same critter will be back tomorrow. He *could* get to be a pest. He's maybe the one Finn Peterson tells about—a male that was raised on a bottle at Turtle Bay."

Now Barney and the female turn toward the open sea and into a mild March westerly. They swim on their backs with their eyes half closed against the bright jewels that glitter from the waves. They are soon beyond the weed zone and over a submarine canyon whose deep slope they have never explored. Each assumes that the other knows where he or she is going. Barney stops to smell a crab-pot float— a brightly painted drum from the bottom of which an anchor line descends.

The female, older and more experienced, drops straight down into purple light. Her sensitive ears have caught the faint crackling of crabs or lobsters in motion. Down, down she urges her body with long strokes of her hind legs, her paws folded on her breast. The sounds grow louder. At fifty-three fathoms, the deepest she has ever dived, she reaches bottom. The pressure on her eyeballs and lungs tells her that she must work fast. More by feel than by sight she pushes her snout against the source of the sounds— a wire-screen box about the size of a bed. She squeezes through an opening, seizes two orange crabs in her arms, and starts to retreat.

But a crab pot is designed to be entered, not left. She

finds her exit blocked at every turn. She drops the crabs and begins to tear in frenzy at the wire. Crimson blood from her lips stains the water. . . . A soft curtain falls across her mind, bubbles rise from her open mouth . . . her life is ended.

When the crabbers hoist the heavy pot next week they will stare in disbelief at a furry rag and a skeleton dripping with sea lice and threads of white connective tissue, the whole blanketed by living crabs.

Meanwhile, far above, Barney grooms in the sun. He wonders where his companion has gone. Twice he periscopes to look for her, then lets the westerly wind sail his drowsy body back into the surging gardens of kelp.

In Monterey, Penny takes a brief holiday from her artwork to continue her education in biology. She believes that an artist, more fully to appreciate the beauty of living things, should understand their architecture and how they "work." So tonight she and Finn Peterson are listening to a talk at the monthly meeting of Otters Alive by a woman known for her studies of respiratory physiology. She is talking on "The Sea Otter as Aquanaut."

"When a man dives deeply in the sea," the expert explains, "the pressure of the water forces air out of his lungs and into solution in his blood. If he rises too quickly to the surface, the air escapes from his blood in fizzy bubbles, as from soda pop. The result is itching, intense pain (known as the bends), paralysis, or even death.

"I've learned from talking to friends in the Divers Club that sea otters usually feed at depths between forty and eighty feet, and that they rise from the bottom quickly, without pausing to decompress—that is, to allow air in the blood to escape gradually. Being curious to know how the otter does this, I studied the anatomy of its lungs. The Fishery Agency gave me the bodies of three animals found freshly dead.

"I was surprised at what I found. The lungs of a full-grown otter are proportionately the largest of any known mammalian lungs—five to eight times the size I would have predicted. I inflated the lungs of an old male to simulate their shape as they would be in life just after a deep inspiration. They measured two feet long!

"You're perhaps wondering why the lungs of a sea otter should be proportionately larger than those of seals and whales, when seals and whales are the more efficient divers. The Weddell seal, for example, can dive under floating ice to depths of nearly two thousand feet and stay down for seventy minutes. And there's a record of a sperm whale diving to *ten thousand* feet and staying down for eighty minutes. I suspect that the depth of a sperm whale's dive is limited only by the time it takes him to get down and back."

She chooses her words carefully. "To understand the large lungs of the sea otter I think we must look at the animal's unique feeding habits. Here's an animal that spends the better part of its life floating, often with a heavy rock on its chest. When it's not floating, it's swimming or diving. Its body represents a compromise between opposing evolutionary forces that have been working, on the one hand to produce large lungs—with buoyancy value to a floater—and on the other, small lungs—with ballast value to a diver.

"Think of the sea otter as a cat gone to sea. Early on it was faced with the problem of storing oxygen during dives. It could have solved the problem either by increasing the

storage capacity of its blood, as did the Weddell seal, or
that of its lungs. It chose the latter course; one can't really
say why.

"The sea otter, although descended from some unknown otterlike beast that would have lived in streams and
estuaries, does not today enter fresh water, even temporarily. This seems odd until one remembers the hard-
pressed economy of the otter. It must float to survive. It
must float while eating, and while nursing and grooming
its pup. Because salt water is more buoyant than fresh, the
ocean offers to the otter a slight advantage in terms of
energy saved.

"I've wondered—and I can't answer my own question
—how a small pup manages when its frightened mother
dives with the youngster in her teeth. I suppose the pup has
to make do with whatever air it happens to have in its lungs
at the moment. And perhaps a good many pups *do* drown."

"You spoke a moment ago of the bends," Finn prompts
the speaker.

"Yes. There are two main reasons why a sea otter
doesn't get the bends, or decompression sickness. In the first
place, it has to contend with only the one lungful of air it
inhaled before diving, whereas a human diver is continuously breathing bottled air. Second, an otter stays down
only briefly—one to three minutes—as compared to a diver's
stay of up to an hour. Again, the small airways of a sea
otter's lungs are heavily reinforced with cartilage and,
since the airways of a *land* otter are *not* so structured, I take
this as evidence that reinforcement enables a sea otter to rise
quickly from deep water. Also, a sea otter's blood has un-
usually high buffering capacity—a chemical feature that

enables the body tissues to recover quickly after breath-holding."

The questioning turns to diving by humans. The expert smiles.

"Each of us started life as a diver. We started as a subaquatic organism in the amniotic fluid, and only at birth learned to breathe air.

"I suppose the most famous human divers in the world are the *ama,* the Japanese women who dive for shellfish and seaweeds. They work almost naked in the cool sea for three hours a day, reaching depths of a hundred feet while breath-holding for three minutes. That's a long time. Try it yourself sometime.

"Recent studies in England, France, and America suggest that a human diver under special conditions could work more than a mile deep in the ocean. He or she would breathe a special air—say a helium-nitrogen-oxygen or a neon-nitrogen-oxygen mixture—and would decompress very, very slowly over a period of a day or so. Already men have dived to 2,132 feet, while a laboratory goat has been subjected in a testing chamber to a pressure upon its body corresponding to 3,600 feet of seawater."

A man in uniform raises his hand. "The navy would be interested in a device—a sort of gill—that would enable a person to breathe the oxygen dissolved in the surrounding seawater. There have been a few experiments in which test animals were immersed in liquid fluorocarbon. A dog, for example, breathed the liquid directly into its lungs for eight hours and then, after its lungs had been drained, recovered and resumed air breathing."

The speaker of the evening answers. "Yes. I know

about that experiment. It was impractical and foolish. The muscles of the human lung are so weak that, unless a diver were breathing compressed air, he couldn't possibly dive more than six feet below the surface of the sea. Below that level his lungs couldn't expand against the counterpressure of the water. I'm saying that, even if he could learn to breathe water, his activities would be confined to near-surface levels.

"I can visualize," she continues, "a sort of mechanical fish gill that would utilize the oxygen dissolved in water—a device perhaps like an automobile radiator pushed through the sea—but it would be heavy and awkward to handle. It would require not only large gas-exchange plates, but a very large heat source, for it would continuously be pushing its plates against water much colder than human blood."

As the speaker carries her lecture forward, Penny soon finds herself lost in technical language. Carbon dioxide tolerance . . . lactic acid levels in the tissues . . . bradycardia . . . the difference between drowning and suffocation. Leaving the lecture hall afterward, she turns reflectively to Finn.

"You know, much of what she said was over my head, but while she was talking I had a bright idea. Nature works like an artist, always inventing, always trying to improve on the designs of her birds and bugs and beasts. She seems never to reach the limits of invention. I hope to be reincarnated a million years from now to see what life is like. And," she adds, "to see what art for the sake of art is like. Maybe it will be wholly invisible . . . electronic waves in our heads, or whatever."

Hand in hand, the two pass into the night. The stars have never seemed brighter.

April

It is April, and the wind off the land is a warm breath laden with the odors of chaparral—the spices of scrub oak, chamise, and ceanothus. Thin clouds veil the sun. Barney floats idly on his back watching the wild geese pass northward to their nesting grounds. They travel in ragged *V*s, now forming, now breaking, and now reforming their lines, each bird winging alone, yet a part of the group. They murmur contentedly. Three brown pelicans materialize from the south, their wing-tips nearly touching the water as they cut the air. Barney settles his body for sleep. He brings his floppy feet forward and folds them over his belly, drops his chin to his chest, and places his paws over his eyes. Within moments he is a graven image adrift on the sea.

Barney's understanding of time—of the days, seasons, and years—is abysmally crude. He cannot know that each new year will bring its seasons in sure and irreversible procession. Yet doubtless he has a primitive sense of before-and-afterness governed by an electro-chemical clock situated somewhere in his brain. At any rate, all marine organisms

thus far explored for evidence of a clock have been found to possess one. Organisms as low in nature's rank as diatoms will jet-propel themselves up and down among grains of sand in harmony with the tides and, long after they have been moved to a laboratory tank, will continue to follow the music of those tides. Sand hoppers (or beach fleas), green crabs, and fiddler crabs, among other sea creatures, seem also to be triggered into rhythmic action by clocks within their bodies.

If Barney *were* to follow a calendar it would be marked by ten thousand dates—by the appearance of gray whales passing south to Mexico in December and north to Alaska in March; by the departure of California gulls in February and their return from inland breeding lakes in August; by the emergence of squid egg-masses (acres of pearly clusters) in March and November; by the sporulating of seaweeds (germs of life that drift like smoke in their appointed seasons); and by the blooming and fading, sprouting and decaying, mating and dying, immigrating and emigrating, of the countless other species that populate his aquatic realm.

So he floats in the sun. When he feels like doing a thing he does it. His assurance of survival rests on his ability to predict, barely seconds in advance, the likely consequences of a selected choice of action.

Now a puff of wind heavy with the smell of a female in heat rouses him from sleep. Forgotten is the pain of his response to that smell in February. He looks around, then starts to cruise toward three otters resting apart from a larger raft. Fortunately, his only potential rival is a male patrolling the borders of the raft and splashing loudly to enforce his territorial claim, unaware that three of his females have defected. Barney approaches the three; smells the tail of the first and then of the second. When a pungent message hits his nostrils he turns abruptly to the third and begins to circle her in growing excitement. She is a sleek, forty-pound, four-year-old who has just ovulated for the first time.

Barney tries immediately to mount her, but she rolls and threatens him with snapping jaws. He persists, churning the water into foam. She bites him sharply on the neck, forcing him to retreat, not quite sure of himself. Nor has *she* read the scenario; she sees only another otter trying to pick a fight. The two spar for a minute. Now Barney advances, pushes her roughly into a tangle of kelp and, astride her belly, seizes her nose and upper jaw firmly in his teeth. He twists her into backside-up position. The rude courtship continues, accompanied by hisses, growls, and whistles. Still grasping her face, his body arching high, he straddles her dog-fashion, places his paws firmly in her armpits, and penetrates her. She folds her paws against her chest; she grows quiet.

As the two curl and jerk in the sea he paddles spasmodically with one hind foot, carrying them both in a wide circle. His other foot is a black leather flag waving in the air in synchrony with his thrusts. At times the pair are wholly submerged, and when the female reappears she gives a gurgling gasp.

On and on, as though struggling to the death, they twist in the water. After fifteen minutes the female sharply contracts her spine and disengages her rear end from the male's penis, though not her snout from his jaws. For a moment the two animals spin rapidly like a beast with two bodies. Then again she grows quiet and the mating act enters its final ten-minute round.

When at last they uncouple, Barney is exceedingly hot. He drifts in the breeze, waving his expanded hind feet in the air. The female's forehead is bleeding, one eye is nearly shut, and her nosepad is a map of red lines. Now the two begin to groom themselves, the female voicing from time to time soft chuckling or chittering notes. When the two are clean they fall asleep, sharing a common couch in the kelp.

At this point, within the warm body of the female an egg is about to be fertilized. The product will be a one-celled zygote, the first building block of a structure that, when complete, will be called a life—an individual—a sea otter. The zygote will divide and subdivide until it is a pearly sphere, or blastocyst, visible to the naked eye, when its growth will suspend. Then, after four months, the sphere will quicken, resume its growth, and reach newborn size after a total gestation of eight months.

This, at least, is what biologists believe. The embryonic unfolding of the sea otter is still imperfectly known. What causes the blastocyst to quicken? Surely not the changing length of daylight, the agent that regulates the cycles of many wild animals, for the sea otter has no one breeding season. It mates in any month of the year.

Also mysterious is the length of the period between the mating act and the quickening of the blastocyst, known as the period of delayed implantation. It varies widely among mammals, being measured, for example, in months in the European land otter, but in days only in the American land otter.

To measure it in the sea otter one would need to hold males and females in captivity under controlled conditions. After each female had been seen mating, she would be isolated from the male and held for a planned number of weeks, when she would be killed and examined. Through autopsy of a series of such females one could reconstruct the early life of the typical embryo. Research of this kind has never been carried out, partly for humane reasons and partly because it would be very costly. To live-capture and feed ten to twenty otters for half a year would require many thousands of dollars.

In the middle of the night Barney wakes, touched by quick foreboding . . . lifts his head and shoulders . . . slowly treads water . . . scans the horizon. His lips taste the wind. Seaweeds draped across his flanks lift and fall. Not fully satisfied, he slides back into his watery bed. He dimly grasps, but cannot understand, that his otter world has been perturbed by an unusual force. Minutes later, the tension in his body drops as an earthquake races along the Calaveras Fault.

What, indeed, is the nature of that soundless invisible warning that, according to country folk and keepers of pet stores and zoos, tells of a coming quake? Some scientists believe that it can only be electrical—that it cuts across the faint individual fields of force possessed by all living things. Others hold that it is simply a geologic signal comparable to the creaking of a wooden beam shortly before it breaks. It is a tremor generated at the fault minutes or hours before a major shifting of the Earth's crust. If so, it is the kind of signal that Barney's ordinary sensors would be capable of receiving, though not interpreting.

At Clam Bay on the evening of April tenth an abalone diver ties his boat to the dock and begins to unload his briny catch.

"I saw some streaks of red water out there today," he remarks to a fish warden who is checking his catch. "Hope she ain't building up to a red tide."

So, when the warden has a free moment he phones his supervisor, for a red tide is often the forerunner of an outbreak of dreaded paralytic shellfish poisoning, PSP. The supervisor reacts by ordering samples of the local clams, mussels, and abalones tested for PSP levels. The test is a painful one. It measures the time spent by a white mouse in struggling to stay alive after being injected with juices from a poisonous mollusk. The Clam Bay samples, fortunately, prove to contain only harmless traces of toxin.

Red tides are patches of water stained with the soft rusty pinks and buffy oranges one sees in Turkish rugs. The colors are imparted by quadrillions of one-celled organisms, classified between plants and animals, called dinoflagellates. Always present in the ocean, they commence, under certain mysterious conditions, to "bloom" or multiply explosively. Plankton-feeding shellfish may ingest the dinoflagellates in such enormous numbers as to become poisonous, not, strangely, to themselves but to many other animals, including people, that happen to eat them. In birds and mammals the toxin acts by paralyzing the nerves that control breathing; it has no antidote and can be fatal.

News of the red water at Clam Bay travels to Monterey, where it provokes discussion at a meeting of Otters Alive.

"I've invited Amelia Caring from the Marine Station to be with us this evening—and I think most of you know Finn Peterson of the Fishery Agency," announces the chairperson. "We're looking forward to learning about the danger, if any, posed by red tides to sea otters."

Amelia responds. "Let me say at the start that red tides are caused by poorly understood chemical and biological agents interacting in the ocean. Many fundamental questions about the tides remain to be answered. Why, for example, should the organisms—dinoflagellates of several species—that are responsible for PSP manufacture toxins when those toxins are of no apparent use to the organisms? They have, at least, no apparent defensive value, for a clam that's feeding on dinoflagellates will continue blindly to feed on them, poison or no poison. Again, why do some red tides bring PSP while others don't? Why do some animals seem to be immune to the poison while others are killed by it?

"Now to the main question: Are sea otters susceptible to PSP? A typical fifty-pound otter eating ten pounds of shellfish a day during a red tide bloom would ingest enough poison to kill a hundred or even a thousand men. Moreover, since the poison affects the respiratory system, and since an otter is especially dependent on its breath-holding ability, one would think that otters would be especially vulnerable to PSP."

Finn interrupts. "If I may comment on your last point, we in fishery research have never seen a mass dying of otters during a red tide. I personally believe that Enhydra is genetically immune to the poison, as are the clams that store the stuff in their bodies without suffering harm. If my theory is good, the otter is like those squirrels that are able to feed on mushrooms poisonous to man. I've seen this myself—a gray squirrel perched on a limb eating a deadly amanita."

"An alternate theory," Amelia counters, "is that when young otters are exposed to PSP at some time or another they *do* get sick but recover and are thereafter immune."

Penny Moreno, in the audience, raises her hand. "I can vouch for the fact that death from PSP is painful. I worked several summers with the veterinarian at Turtle Bay. A client brought in a housecat that had been scrounging on the beach during a red tide. It died gasping for life on the examination table while we were trying to save it. Another time, a fur farmer brought in a dozen minks that had died soon after being fed topsmelt. The client was puzzled; he said the fish had appeared to be fresh and wholesome. We weren't able at the time to pinpoint the cause of death, but later we heard on the radio that all the beaches along our part of the coast had been closed to clammers on account of a red tide."

The discussion goes on. Can a red tide be predicted from weather forecasts? Evidently not, although dinoflagellate blooms are most frequent during warm, calm spells in spring and fall. They often follow heavy rains and floods when higher than usual concentrations of nutrients reach the surface water. Can an individual clam be identified as poisonous without sacrificing the life of a laboratory animal? Not at present.

"Most of us will agree," concludes the chairperson, "that a simple analytical method for detecting PSP ought to be developed. And, when it *is* developed, the Fishery Agency should routinely apply it to the stomach contents of sea otters found dead during red tides. Perhaps then we will get the answers to some of the questions raised here this evening."

On a morning late in April in Otter Cove a strange performance is being played. Barney sees it from a distance but pays no attention to it. It is being played by six employees of the Fishery Agency and a news reporter. A woman equipped with binoculars and a walkie-talkie radio sits on a high rock at the margin of the cove. The others, two of them dressed in wet suits, are cruising slowly in a motorboat. A second boat follows. Finn Peterson, in the leading boat, talks to the reporter.

"Our experience so far with tangle nets has taught us that we can catch otters, but it has also taught us that we can expect losses from strangulation, shock, or exposure. The rig we're going to try today is a Rube Goldberg invention and it may not work, but we're hopeful. We call it a diver-held trap. I'm sure there's nothing like it elsewhere in the world." His hand drops to a device about eight feet long—a pole ending in a barrel-shaped net, the whole construction resembling a ship's mast topped by a crow's nest.

The reporter smiles. "Make a better mousetrap and the world will make a beaten path to your door." He tactfully refrains, however, from airing his private opinion of the device. Anyway, what a fine day to be out on the blue Pacific.

Finn continues. "Our agency is planning to move large numbers of otters—maybe hundreds—from places along the coast where they're abundant to places where they have not reestablished themselves through natural immigration. Our relocation program has two goals: to quiet the ongoing argument between clammers and otter lovers and to ensure the survival of the species in the event that catastrophe, such as a crude-oil spill or a nuclear blowout, strikes the central California coast."

"I know what you mean. I remember the mess at Santa Barbara ten, fifteen years ago. Miles of beaches covered with tarry oil. Dead birds by the thousands. Lots of well-oiled seals. I don't remember seeing any dead otters, though."

"No, the otters hadn't reached that part of the coast

yet." The boat chugs along. Finn lowers his voice, for he sees ahead what may be an otter.

"I should have added that our relocation program has a third goal—to capture a few animals for use in research and education. In the past, when otters were scarce in California, we had a firm policy against taking any dead or alive. Now we feel that a few animals removed from the population each year to increase public understanding of, and sympathy for, sea otters is a proper use of the species."

". . . two hundred yards northeast of you . . . fast asleep . . . a big one." A message from the woman on shore crackles in the boat's receiver.

The boat stops. "Take her away, men!" Two husky divers who have rehearsed this bit slide over the side and take the trap in tow. Glancing at the sun, they head northeast, swimming underwater, rising now and then to set their course on a dark shape bobbing in the distance. The otter, a full-grown female, feels the faint vibration of the trap as it parts the water. She raises her head but, seeing only a boat—a familiar sight—falls back in repose.

. . . Five minutes pass. The men are now directly beneath the otter; her silhouette shows plainly against the sky. Carefully one of them turns the pole to vertical, then thrusts it upward in a powerful stroke. The rim of the net breaks the surface and surrounds her. She dives in a flash but finds her body blocked by the webbing, allowing the men time to pull the purse string of the trap and to jerk the valves of two floats attached to it. These inflate, hissing, with carbon dioxide. Then the men back away.

The otter tears angrily at the thing, which seems a weed of unusual toughness. Reflexively she screams a warning to others of her group; it is thrown back by the shore.

Meanwhile the powerboat has reached the scene of action. Finn and his helpers lift the otter over the gunwale. As though she understood that her inborn defenses were now useless, she huddles quietly on the floorboards, making no attempt to bite her captors. Her brain can suggest no move, no plan of action.

"Con-grat-ulations!" comes the voice of the woman on shore.

"Nothing to it," replies one of the divers. Then, eyes shining, he speaks to his buddies. "I was real surprised that she didn't take off. She must have known that something was underneath her. I suppose she was only half awake and thought it was another otter. If we can catch them this easy we're in business!"

With gloved hands, Finn lifts the otter from the trap and transfers her to a cage in the second boat, then directs the search for other sleeping individuals. Before sunset the team will take two others without injury to the catchers or the caught.

One by one the captives are released in a floating pen, where they will remain until all danger of losing them from shock has passed. The reporter returns three days later to witness the final act of the performance—the transfer of two otters (one having died) to San Pedro Island, a place far to the south where otters flourished in the nineteenth century before they were hunted to extinction. Finn reports on the welfare of the captives.

"Our veterinarian looked at the animals. Their blood sugar is a bit low but it's no cause for worry. He says we can expect it in animals under stress. The little female that died yesterday had gut inflammation, an ailment she may have had in a mild stage when she was captured. We've weighed and tagged the other animals." He indicates a floating otter wearing, on the right foot, a yellow tag and on the left a red, white, and blue one—both numbered 35.

"We've improved our tagging method. We can now identify a colorcoded otter from a distance of three hundred meters. We've also tried peroxide and other cosmetic bleaches to print a conspicuous mark on an otter's back, but they're unsatisfactory. Too slow and messy.

"You spoke of stress," pursues the reporter. "Can't you tranquilize them? Give them a pill or a shot?"

"We've tried it, but it's tricky. The basic features of our capture-and-release method were worked out in Alaska some years ago. If memory serves, the first attempt to capture otters alive was made in 1951. All the animals died. Drug injection has been used to quiet very aggressive animals, but it throws off their thermostats, so to speak, and can lead to chilling and death. We'd rather handle the animals gently, keep them in clean running water, and feed them the kinds of food they like."

"I must say," offers the reporter, "that your animals look healthy. I'm surprised that they tamed so quickly." As if he understood the compliment, a white-headed male corkscrews on the long axis of his body, plunges to the bottom, and rises in a fountain of spray.

"What's that one eating?" He points to a female chewing on a floppy gray object.

Finn chuckles. "A rawhide toy for dogs. I got the biggest one I could find. It works as a pacifier. The otters spend considerable time chewing it and whacking it against the poles of the pen." He turns his face abruptly to the sky.

A small floatplane drops to the surface of the cove and glides to a halt with its prow on the beach. The agency workers are ready with two plywood boxes lined with damp sponges. They stow the animals, one to each box, in the cargo compartment of the plane and an hour later release them on a sandy beach at San Pedro Island. Here the displaced animals will find abalones, clams, urchins, and lobsters in abundance.

Two men from The Network are on hand to watch the release. They, and others who will replace them, plan to monitor the beachers every few days and to report on the success or failure of the transplant.

Meanwhile two other men, commercial abalone divers, are watching from a boat anchored offshore. It is well that their salty comments do not carry to the ears of those on the beach.

May

White moonlight falls on a raft of otters off Silica Beach. Numbering about a hundred, held loosely together by imaginary cords, they touch and drift apart in the tidal currents. They talk softly in gargles, chuckles, and whistles. Now a pup searching for its mother gives voice to a thin complaint and now a small fish pursued by a larger one jumps from the water and reenters with a *plop*. Why the otters are gathered here in such great numbers is a mystery, although doubtless they are following some blueprint designed to perpetuate their species. They constantly groom yet, despite their social nature, do not groom one another, except mothers their pups.

Barney and his mate, at the center of the raft, rise and fall with the sea, wholly absorbed in whatever they are doing in the night or are thinking of doing. Because the hairs of her face are tipped with white, and because she will return often to the pages of the present story, she may be called Paleface. Barney watches absently as a large pup sneaks up on a sleeping grandfather, grabs him by the base of the tail as one would grasp the handle of a jug, and pulls him underwater. What fun! The old fellow snorts. Were he not so drowsy he would box the pup's ears.

A white shark as long as an automobile, attracted by
the submarine noises from the otter raft, slides quietly in
from the ocean. In the first light of dawn his lead-gray form
is part of the lead-gray sea. His black eyes express a fixity
of purpose sharpened by two hundred million years of
practice in the art of search-and-destroy. As he gathers speed
his entire body begins to undulate—a motion the more
awful in its utter silence.

The last image of the world registered by a certain
mother otter is a circle of gleaming, razor-sharp teeth
growing by the instant . . . then oblivion. Nearly torn in
two, her body twists like a falling leaf while the shark circles
to hit her again. A pool of blood reddens the surface. Un-
hurried, the shark returns and tears at her forty-pound body.
Tatters of skin and fur float up to be fought over by gulls.
But a shark's teeth are better adapted for cutting and sawing

smooth-bodied fishes and seals than for dissecting thickly furred otters. The shark tugs at the warm body for a while and then, not being especially hungry, leaves it.

He gathers momentum to pursue a half-grown male fleeing in terror toward shallow water. The otter's speed is no match for that of the hunter, who is moving at nearly twenty miles an hour. An instant before impact the otter twists to one side, thereby saving his life but losing his tail. He will swim erratically, like a rudderless boat, for the rest of his days. At last the great fish, primordial enemy of life in the sea, melts into the gray infinity from whence he came.

The body of the first victim will be found tomorrow on Silica Beach by a volunteer from The Network. He will carry the damp thing in a gunny sack to a veterinarian, who will discover a clean semicircular gash exposing heart and lungs. The vet will find a triangular tooth embedded in a rib and will mail the tooth to the curator of fishes at California Institute, who will reply that it came from a great white shark, or maneater, about fourteen feet long.

At this point our narrative veers away to report an incident that begins in a tavern at Turtle Bay in late May. Three abalone divers, bearded, wearing sweatshirts, are talking themselves into a nasty mood. Soon they leave the tavern, load a mysterious package into a motor launch, hide the boat's identifying number under a tarp, and put out to sea. They clear the breakwater and cruise northward to a place beyond sight of anyone on shore—or so they think. In the hazy midday hours they break out two guns and begin shooting into a kelp bed. They reel and dip and catch themselves on the heaving deck.

They cannot know that Finn Peterson has chosen this very day to take a holiday. He is perched on a cliff, sandwich in pocket, 60-power telescope trained on a raft of otters in the kelp. He is timing the dives of a certain conspicuously marked, black-furred individual. The Coast Guard chart shows ten fathoms here, and the otter has been monotonously diving to the bottom every three or four minutes for over an hour.

Suddenly the water is pocked by small fountains of spray; the sound of gunfire reaches Finn; the otters start to rise or dive in alarm. He focuses on the boat and mutters, "What the hell's going on!" He sees the men hauling two brown objects from the water. He grabs his telescope and scrambles along a goat trail to the highway and his car. Twenty minutes later he is talking by phone to the local warden.

The shooters see him moving along the face of the cliff and, as they near Turtle Bay on their return, they see from a distance a blue-and-white Fishery Agency van parked on the dock. They hastily push the brown objects over the side and, after a brief argument, the rifles as well.

On the dock they are arrested by two wardens for shooting at a protected species. "We're *clean*," they protest. "Search our boat if you want to, but you'll be laying yourselves open to false arrest." Nonetheless, the wardens search the boat, taking as evidence a handful of shell casings. One of the pair dons a diving suit and drops to the bottom, where he had seen the suspects throwing something into the bay. He finds a Marlin rifle, while from the surface he recovers two dead otters. Back on the dock he opens their bodies and records the fact that their viscera are still warm to the touch.

Some weeks later, the agency makes its case in criminal court, where Finn presents his eyewitness evidence. Slugs found in the bodies of the otters match the rifling of the Marlin. The otters had apparently been dead less than an hour. The clincher comes from a Turtle Bay merchant, who testifies that one of the accused bought five hundred rounds of Marlin cartridges on the morning he was arrested and, later, tried unsuccessfully to have the sale slip destroyed.

"I fine you each one thousand dollars," says the judge to the three men in court. "And speaking as a private person, I fail to understand how you could have killed a sea otter. Who are you to say that *your* proprietary right in abalones is greater than *mine* in otters?"

The scene now shifts to the pier of the Marine Station, where on sunny days Amelia often eats her lunch. Here she watches the shore vegetation come in and out of bloom and watches the moving parade of seadrift in the water at her feet. Here she tries to interpret the behavior of the birds flying above her head. Today, listening to the broken cries of a sea otter somewhere out in the bay, she wonders, *Why not make a tape recording of otter sounds?* At the outset it would have no practical value but, if one were gradually to become familiar—through direct observation—with the meaning of each sound, the recordings could become a sort of sea otter glossary. Already she feels certain that the voice she hears is a distress cry and that it comes from a pup.

She absently tosses a crust to Old Charlie, the one-legged California gull who has long claimed the station as his private bird preserve.

That afternoon she talks with a retired scientist, a specialist in underwater sound who acquired his skills in

World War II when he was assigned the job of distinguishing between, on the one hand, the sounds of ships and submarines and, on the other, the natural sounds of shrimps, fishes, seals, dolphins, and whales.

"If you plan to record underwater sound," offers the specialist, "the apparatus is bulky and expensive and you risk losing it at sea. Last year I loaned some gear to a young fellow who proposed to install it in a kayak so that he could move in close to gray whales and record their faintest signals. Well, he did, but a whale rubbed his back on the kayak and tipped it over. Now my gear is lying on the bottom of Magdalena Bay. But *open-air* recording is fairly easy. I'll be glad to help you with that. I have a tape recorder and a big ear. All we'll need is fresh batteries and tape—and, of course, a boat."

"A big ear?" Amelia raises a brow.

"A one-meter parabolic reflector to bring in distant sounds."

So, on the next fair Sunday morning two boats might be seen moving through the blue water beyond the kelp beds of Monterey Bay. A motor launch with five passengers is towing an inflatable rubber boat. In the launch are the specialist, Amelia (with binoculars bouncing on her bosom), two volunteers from The Network, and the owner of the launch. When they have come within two hundred yards of an otter raft they don life vests and transfer to the rubber boat. As they paddle soundlessly toward the raft, the otter small talk grows plainer. Now they will drift for two hours on the current, recording each new sound.

Baaaw! Baaaw! . . . "A pup," says Amelia softly to the volunteer who is acting as note-taker. Binoculars glued to her face, Amelia continues, "Now mama is surfacing beside it with something red in her paws. She drops it on her chest and licks the pup's face. The pup quits bawling." The volunteer glances at her watch and notes the exact time so that the observation can be synchronized with the tape. She scribbles in her notebook.

An earsplitting scream. "An adult standing upright in the water . . . mouth wide open. Can't tell its sex or why it's screaming."

The specialist removes his earphones. "That was a beauty. It would have carried half a mile downwind."

Whee! Whee! Amelia shakes her head doubtfully. "A shrill whistle near the upper limit of my hearing. It's coming either from a subadult or an adult. . . . They're facing each other. . . . They seem to be arguing."

The otters in the raft pay less and less attention to the boat as it drifts among them. The sun climbs toward the zenith. The passengers drip with sweat beneath their vests. Suddenly they see a bloody-nosed individual attended by a larger one.

Ku-ku-ku! Amelia reports: "A female . . . just mated, I think . . . one of a pair. Her face is red and bitten. He's circling her now. I think it's she who's making the cooing sound, although her mouth is closed." The sound is repeated. (Later it will be recognized as an expression of general satisfaction, for it is heard also at times when an otter is eating a tasty food.)

The specialist raises his hand for attention and plays back a section of tape. "What's this?"—a sound like croutons being crushed in a bowl.

Amelia smiles. "Breaking shells. An otter's teeth can crack any crab or lobster, and most clams or snails."

And so the morning goes. Amelia frowns as she tries to

translate animal vocalizations—which are, of course, untranslatable—into words such as snarl, growl, hiss, grunt, or bark. Then she thinks, *It doesn't matter; they're all on tape anyway.*

A volunteer points. "Just beyond that white rock there's an otter with a nose too big for his face—a real Durante schnozzle."

Amelia targets the animal with her glasses. "I see it now . . . a large pup. An otter's nose grows faster than its body. The nose of any breath-holding animal has to be well developed early in life. This one's short whiskers tell us that it's a pup; they're not half as long as they *will* be."

She stares for a minute, then chuckles. "I was watching the little one beyond it—a pup about two weeks old. It seemed to be swimming like an adult, when I saw that its mother was moving full-steam ahead underwater, ferrying the little bugger on her chest."

A thin cloud moves across the sky and wavelets begin to whisper at the sides of the boat. "Time to quit," announces the specialist. "We're getting too much surface noise. But we've taped some really first-class music. I had no idea that otters would be so expressive."

In the afternoon the adventurers gather in the library of the Marine Station to swap ideas. The specialist begins.

"I understand that you folks are basically interested in message meaning. Don't be overly optimistic. As you gradually decode the various otter signals you'll find them quite simple, having the barest of meanings. Incidentally, when I speak of sounds produced by nonhumans I use the word *signals* in preference to *voices* or *language*. Our next step will be to catalog the sounds we captured today and to prepare a sonogram, or visual printout, for each.

"From past experience I would guess that most of the sounds we heard were concentrated between three and five kilohertz, or well below the highest pitch—around twenty thousand—audible to the human ear. And I think the printouts will show us that the mother and pup were *exchanging* signals. Resorting to dialog would help them keep in touch with one another, especially at night and in fog."

Amelia changes the subject. "How would you go about testing sea otter *hearing?* I know from experience that an otter can hear and locate a movie camera running in the open air at a distance of at least a hundred yards."

"A good question. We could test it by playing back recorded otter sounds. While watching an individual animal from a blind on shore—or from the station pier—we could gradually increase the sound volume until we noticed a response. We could then compare the hearing sensitivity with that of, say, a dog or a man."

A volunteer speaks up. "It can be argued, I suppose, that because sea otters live in a narrow world, they have little need for acute submarine hearing—and if they don't need the hearing they're unlikely to have it. By narrow I mean that their environment is a coastal strip in which they move up and down and sideways but seldom out to sea."

"An interesting point. And *I've* often wondered whether the sounds of civilization—of human traffic—may not be affecting the nervous systems of otters and other marine mammals. The average otter must hear hundreds of sounds from foghorns, ships, and aircraft, as well as explosions of many kinds, all totally foreign to an otter's ear. Are these spooking him? Disrupting his finely tuned feeding and breeding habits? Or, on the other hand, does an otter become desensitized to man-made sounds early in life, even in fetal life? I don't know."

He continues. "I was once asked by the Fishery Agency whether a sonic barrier would keep sea otters away from commercial shellfish waters. I had to express doubt. Any 'scare-sound' device would be costly to maintain in the open sea and, more important, the otters would probably get used to it and would ignore it."

"As a matter of fact," Amelia says, "I've researched that point. Alaskans tried to scare otters with helicopters, firecrackers, rockets, and carbide cannons. The few animals that fled the scene returned shortly after the racket stopped.

"Well," she says, rising. "I don't see how anyone could measure the acoustic damage to an otter's feelings as a result of man's noise. Anyway, I'm sure that today's experience will start us thinking about otter hearing and speech—pardon the expression—and will stimulate further ideas about research design." She turns to the specialist. "We're very grateful for the insights you've given us into the otter's world of sound."

The month of May along the central California coast is a time of transition, a time when the hills have burned to browns and yellows, while the shores, daily bathed in morning fog, have kept many of their lively greens. May is the month the airs of Barney's world are ripe with the smells of nesting seafowl. These are skatole smells—fishy, pervasive, stubborn.

They rise from the nests of cormorants, rude mounds of seaweed closely packed on the ledges of coastal cliffs—

and from the nests of oyster-catchers, bowls lined with pebbles and fragments of crab claws and shells, built in the open near high-tide line

and from the nests of western gulls (the only gulls that breed in Barney's territory), built of grass on the slopes of offshore islets

and from the nests of murres, mere splotches graced by a naked egg, on rocky plateaus and ledges

and from the nests of pigeon guillemots, casually built on bare rock at the bottom of some crevice, or at the end of a tunnel in an earth bank, or among drift logs.

Toward the end of a starlit night, squids rise to the upper waters in shoals of thousands. They have come from deep water to mate in the shallows. Their frenzied movements induce vibrations in the water, which Barney reads as an unusual event. He wakes, rolls over, and stares down at pearly shapes, pale torpedoes darting in all directions, swimming by jet propulsion. Their enormous silver-and-black eyes pop with excitement. Now a male meets a female, reverses direction, and grasps her around the middle with his tentacles. Immediately two other males cut in but are repelled after a bumping skirmish. Another female, pursued by a male, shoots upward into the night with an audible *whoosh* and falls on Barney's rump.

This is too much. He begins to bite and gulp down one fat, soft creature after another. They are helpless, for the mating urge has short-circuited their normal instinct to flee the hunter. He eats as never before.

The pink flush of dawn touches a furry belly swollen by two quarts of squid as Barney returns to sleep among other gluttons, or opportunists, of his kind.

June

On this June morning, unbeknownst to Barney, he is five years old.

No sooner has the mist risen from the sea than he hears the pulsing din of an airplane coming from the south, traveling no more than a hundred feet above the water. Its shadow darkens his head at the moment he dives in fright. Paleface takes evasive action by spurting full-speed ahead, then leaping in the air like a salmon. A startled mother seizes her pup in her teeth and disappears beneath the surface. A subadult climbs on the back of another for a better look. The annual June census of sea otters along the California coast is underway.

Meanwhile, at a meeting of Otters Alive the members are talking about historic changes in California's otter population.

Finn Peterson pulls thoughtfully at his Viking beard. "The growth rate seems to be slowing, especially at the middle of the geographic range, where otters have lived for forty years or more. I would guess that the population

is nearing equilibrium in that region. It's close to the carrying capacity of the range.

"In trying to understand the changes in California's otter population I've found it helpful to study the changes in Alaska's larger population. Alaskan biologists have carried out more intensive research on otters than we have, mainly because at one time there was talk of establishing a commercial sea otter fishery up there. In fact, the Alaskans did take around three thousand pelts from 1962 to 1971, but quit when they found they couldn't make a profit.

"Well," he continues, "we're not planning to start shooting or clubbing California otters for their skins. But back to my point—Alaskan biologists estimated that on a certain island long occupied by otters, the population had stopped growing, while on a nearby island long *unoccupied* by otters a new, pioneer population was growing at 10 to 13 percent a year."

A voice interrupts. "I don't quite understand; would you explain, please?"

Finn responds. "On Island A, where food was scarce, the otters' reproductive rate balanced the combined death and emigration rates. On Island B, where food was plentiful, reproductive rates were high and death rates were low. These factors, together with the influx of animals from Island A, brought rapid growth to Island B's colony."

Amelia Caring turns to him. "How accurate are your census methods in California? The figures published by the Fishery Agency seem to jump around considerably from year to year."

"They do, but we're steadily improving our census techniques. We make head counts simultaneously from the air and from shore. We take photographs, using the best aerial cameras we can get. And we adjust the final count for variables such as wind, sunglare, and dense kelp cover, all of which reduce visibility. The best condition is what we call milky water—a flat sea under a thin bright overcast. An average figure for population density would be somewhere around twelve animals per mile of shoreline . . . maybe less.

"And we're starting to estimate the percentage of what we call clinging pups. It's now running 10 to 15 percent of the total population. As the percentage changes in the future it will serve as an index of the fertility of the population."

"But the *accuracy* of the counts," Amelia reminds him.

"Well, I recall a January count made along a strip of the coast in *good* weather that was twice as large as one made a month earlier in *bad* weather. But we make four or five partial counts a year, and when we average them they trace a fairly uniform line."

Penny Moreno's mind is far away. Once again she sees Barney come home to Turtle Bay, standing in the sea and staring sharp-eyed at the passengers of the *Linda*. He is Poseidon in the clothing of a beast. . . . Then her mind snaps back to the real world.

"Finn," she asks, "why is it important to know exactly how many otters there are? I can understand that if one were managing them as livestock one would want to check on the capital investment, so to speak. But they're *not* being exploited and I can't see why the agency has to spend thousands of dollars every year making a head count."

For a long moment Finn searches the rafters of the old warehouse. "The custodian of every national park keeps an inventory of the important wild animals living in the park. Visitors like to be told, for example, that a park is home for ten thousand elk. The visitors take pride in a national wildlife treasure even though it isn't being *used* in the everyday sense of the word. And Californians, by and large, like to know the size of the treasure represented in their otter colony.

"There's also a legal reason for counting. The California otter colony is classified under the Endangered Species Act as a 'threatened population.' We're required by law to keep track of its size.

"Finally, we've learned that year-to-year changes in the numbers of any wild species reflect the condition of its environment. The species can serve as a 'miner's canary.' And when the environment seems to be deteriorating we can sometimes reverse the trend."

"May I return to a point?" asks Amelia. "What is a threatened species?"

"According to law it's any species *likely* to become endangered within the foreseeable future. As you can well imagine, this definition has given rise to opposite opinions— that California's otters are, and that they are not, *likely* to become endangered. Those who hold the first opinion have been called alarmists."

"Then I'm a genuine, card-carrying alarmist," says Amelia quickly, "because I'm certain that California is *likely* to suffer a major oil spill in the foreseeable future. The oil well that exploded off the coast of Mexico a few years ago leaked for ten months. It fouled the Caribbean beaches and waters with millions of barrels of goo. God save us all, and especially the otters, if California should suffer a disaster of the same magnitude. Our time in history will be remembered, I'm afraid, as the Time of Dirty Energy."

"I'm more optimistic," counters Finn, "because I think we're about to get the oil industry under control. I *will* agree that oil and otters don't mix. Our agency funded a study at California Institute in which otters were deliberately painted with oil and returned to their pool. Over a period of a week or so their metabolic rate rose substantially. In one case it more than doubled. The animals were losing heat through their soiled fur and were compensating by stoking their internal furnaces. A soiled animal in the wild would never be able to feed fast enough to keep warm. It would die."

A volunteer raises his hand. "Recent information from The Network will be of interest to you. Last February, off the coast of Big Sur, a storm hit a barge carrying lumber and pushed two million board-feet into the ocean. Since that time we've been keeping track of the arrival of floating lumber on the central California beaches. It now covers nearly all the beaches within the range of the otters. If the movement of floating oil is comparable to that of lumber, an oil spill under similar circumstances would foul the habitat of 90 percent of California's otters."

June days on the central California coast are lovely days. The hills sleep in the sun. Locusts buzz and robber flies drone. The aromatic smells of sage, golden yarrow, and tarweed lie heavy in the air.

Barney drifts on his back, holding an orange starfish in his paws, chewing thoughtfully on one of its tough, grit-studded arms. As though deciding *to hell with it,* he lets the mangled creature sink into the sea. The starfish is not fatally wounded for, like many a primitive organism, it can regenerate parts lost through accident or predation.

He sculls gently toward Paleface by moving his extended feet, soles up, in rhythmic up-and-down undulations. Only her rump is visible, for she is head down in the water nibbling at a cluster of gooseneck barnacles hanging from a floating log. Each barnacle has a brown leathery neck about three inches long surmounted by a head of coarse limy plates. To chew one must be like chewing pumpkin seeds in their shells. Barney's interest in food soon gives way to another. Sex.

If the sexual behavior of male sea otters in the wild is like that in aquariums, they are easily aroused. In captivity they have erections—with penis extruded from furry pouch —in every month of the year. One aquarium keeper saw a male try at least weekly for more than a year to mount a female who shared his pool.

So, Barney grabs Paleface around the neck and pulls her away from the log, mouthing her lightly on top of the head. She twists in annoyance, faces him belly-to-belly, and grabs *him* around the head. She bites him hard on the nose. Jousting for position, the two spin for a moment, when she breaks loose and shoots like an arrow into deep water. Her brown fur enveloped in bubbles is a shining coat of mail. Barney doesn't care to repeat this act—for a while, at least. He lets his body cool in the sea breeze.

Sex aside, there is something touching about the social warmth or comradeship that sea otters display toward one another. They rarely fight, and none has ever been known to kill another. After Arville Whitt had spent two thousand hours watching the behavior of sea otters in aquariums, he concluded that they

> . . . are very dependent on each other as companions and they forcefully resist efforts to be separated from other otters. If one is separated it screams until it is allowed to return to the others [but] if two are isolated from the rest they do not scream.

He observed that when captive otters were frightened, as they were on several occasions when a visitor approached their pool with an open black umbrella, they

> . . . would grab each other around the neck and face the danger together, retreating slowly in the oppo-

site direction. They did not release their holds until
the danger was past.

When a mother otter fell seriously ill she was removed from
the pool for treatment.

> At removal time two keepers took hold of [her] and
> began pulling her out of the water. Her pup locked
> his arms around her waist and pulled frantically,
> trying to keep her in the pool. The tug of war ended
> in a few seconds and the attendants walked toward
> the den carrying the female, while the pup followed
> very closely behind her uttering sounds that the at-
> tendant referred to as crying.

When two otters, either captive or wild, play roughly
and tease one another, neither seems to take offense or
attempt to retaliate by biting. The first to tire simply fades
from the scene. An exception is the violence displayed in
sexual encounters between male and female, or between two
rival males.

In the evolution of Enhydra, nature's rewards con-
sistently went to those individuals who were slightly more
pacific than their companions. An otter's life at sea is at best
a hard and dangerous one. The daily problems of that life
are more easily solved through cooperation, even if it be at
a primitive level, than through individual effort alone. It
is a life in which prolonged infant care and rafting in groups
are clearly important. It was no accident that gentle
Enhydra, within its demanding environment, found survival
value in behaving with tolerance toward individuals of its
own society.

*   *   *

A pink-and-lavender sunset colors the June sky and the waters beneath it. The light is filtered by smoke drifting in a thin veil for a hundred miles out to sea, its origin a chaparral fire behind the Big Sur. Gulls, their mewing stilled, fly one by one to their resting places on offshore rocks. Where the runoff from a salt lagoon meets the sea, a heron stands in the darkening light, its neck a curve on an Egyptian vase.

The sun disappears. Sea and sky become for a moment a seamless curtain of gold.

July & August—
Midsummer

A hot wind from the hills presses down upon the bay where Barney rests. By mid-afternoon the temperature will reach 100 degrees. The sun stands naked in the sky. One might think that Barney would suffer inside the fur coat that covers all his body except snout and palms, and perhaps he does, but he knows how to seek relief. He moves as little as possible, changing position only to wet his fur from time to time. Much as a family dog will lie on its spine to have its belly scratched, Barney lies in the water with his head, belly, limbs, and tail in the air. He breathes in shallow drafts, a trick that leaves residual air in his lungs and enables him to float high in the water. Now and then he ducks beneath the surface to wet his fur. Five to ten times a minute he sweeps the air languidly with a hind foot, beginning each stroke with the leg fully stretched, then hauling it slowly forward, spreading the webbed toes like a fan. On the return stroke he closes the fan. He is a self-contained air-conditioner.

Writing in 1874 about sea otters, Charles Scammon,

sealer and whaler, noted that "they are frequently seen . . . with the hind flippers extended as if catching the breeze to sail or drift before it." Captain Scammon, it would seem, knew much about sailing but little about thermoregulation.

Barney sighs. He lets his body sink through a canopy of kelp into cathedral light . . . his paw touches a frond . . .

he comes to rest in midwater. Here in the sea and of the sea, neither asleep nor awake, he sways with the stems. For a while he does nothing at all.

And meanwhile, what of Barney's old friend and pediatrician, the animal doctor of Turtle Bay?

"I had a long day," says the vet, sinking into a lawn chair with a frosty drink in hand. "A long, long day." He is talking to Penny, home for the Fourth of July weekend. "The marshal got me out of bed at five this morning . . . said there was a sick whale at Slater's Beach. I picked up a couple of fellows at the marina and we drove out there. Found a bottlenose dolphin about ten feet long wallowing in the surf. It looked to be unharmed, so we waded in to our waists and tried to head it out to sea. It insisted on turning around, almost as if it *wanted* to strand itself. Finally a big swell picked all of us up and dumped us on the beach."

"What was wrong with it?"

"We didn't find out. It took its last breath while we were watching it. Maybe pulmonary edema or congestive heart failure. Maybe it had swallowed a sharp object that blocked or punctured its intestine. There's a lot of junk floating on the ocean these days. Gut trouble is fairly common among dolphins—and among seals and sea otters, too."

"Is it still there? I'd love to sketch it."

"I'm sorry. I didn't know you were in town, so I asked the marshal to have it towed out to sea and sunk.

"Well," he continues, stirring his drink. "We started home. When we rounded the bluff at the base of the breakwater we heard what we thought was gunfire, so we stopped. Two kids were dropping firecrackers into the little cove where the otters hang out—you know the place. There were no otters there by the time we arrived. We threw a scare into the kids and took their names, but I don't intend to make a case out of it.

"So, I came in, changed my wet clothes, and ate my breakfast. The marshal phoned again. It seems there's a fellow in town calls himself Captain Stark. He and a lady friend own a trained-seal show on wheels—a big van carrying a collapsible pool and pens for two sea lions, with sleeping quarters topside for the captain and his friend. Yesterday, while the van was in a parking lot, both the animals suddenly died.

"Back to town I went. The captain was a seedy character who looked as though he had crawled along the darker edges of life. The lady was a frazzled-rope blonde— nice enough, though. She was crying. I looked at the animals and was inclined at first to agree with the marshal that they'd been poisoned. You may not know it, but there are screwballs in this world who will secretly poison a captive animal to 'release it from bondage.' Then I looked at the pens and saw that the animals had darn near torn them apart. Strips of wood and wire, shreds of rubber matting—the works."

"I pumped the captain for details. He admitted that he and the lady had left the van for a while yesterday to visit a tavern. But it was 100 degrees in the shade here at noon, and likely 110 or better inside that van. The animals had simply died of heat prostration."

"How awful! How very awful!"

Penny and the vet were never to learn that the captain had lost interest in this particular lady and these particular animals. He had deliberately shut off the ventilator of the van.

The vet sighs. "From then until six this evening I was

busy in the clinic. Should have taken my phone off the hook. Hell, it's a *holiday!*"

Next morning, Penny and her brother hike to a small beach protected from the sea by the basaltic bones of an ancient lava flow—a quiet place that can only be reached by carefully picking one's way down the gutter of a dry stream. Penny carries in her pack a camera, a notebook, and a towel. She plans to photograph things of beauty and interest along the shore. The photos will serve later as memory guides while she sketches and paints.

The two are on the beach at sunup, for this is the time of day when the air is clean, convection winds have not begun to blur the leaves of plants, and the sun is casting those long shadows that give depth and brilliance to a photograph. Unruffled tide pools offer mirror reflections; sand patterns are fresh.

She captures the image of a cockleshell flawless in line and geometrically exact . . . of an abstract design in blue-white sand and blue-gray stone . . . of a root pounded so often by the sea that only its skeleton remains . . . of bedrock carved into sensual curves . . . of the dimpled track of a crab like a space-vehicle track on the moon . . . of the white eroded shell of a sand dollar still bearing its starlike tracery.

The sun moves higher, shedding more heat than light. Penny is ready to quit when her brother approaches, carrying an object in his hands.

"An Indian mortar! I was hunting crabs under boulders when I turned this thing over and saw that it had been hollowed out."

The thing is a solid granite rock about eight inches

across, in the shape of a grinding bowl, obviously worked by human hands. Long scoured by the surf, it still shows attachment scars where oysters had grown while it lay on the bottom. It looks very old.

Penny lifts it. "Beautiful. I've seen them in museums, but never hoped to find one, especially in the water. How do you suppose it got here?"

Tony rubs his fingers over the flowing contours shaped by a long-dead artisan. "I've heard the divers tell about finding them—sometimes a long way from shore. Mortars, and pestles, and net weights."

"Net weights?"

"Rocks shaped like an egg with a groove around the middle to hold a line. This thing may have fallen from an old campsite. The surf is always undercutting the cliffs."

Penny, more imaginative, offers a different explanation. "Well, *I* think it fell from a canoe. A girl was carrying it from one shellfishing camp to another. It was an heirloom of the tribe. Suddenly a great green wave overturned the canoe and drowned her dead . . . and her twin babies with her."

"You've been in the sun too long—let's go." Tony makes a sack of the towel, wraps the mortar in it, hoists it over his shoulder, and starts up the trail.

As Penny sits on a log tying her sneakers, her eye is caught by a curious object half-buried in drift. She picks it up and see that it is a skull the size of a small gourd. She says to herself, *Sometime I will draw it.* She shakes the sand from its chambers and drops it into her pack.

Next day, she stops by the veterinarian's to say good-bye and she casually mentions the skull.

"May I see it? I've started a skeleton museum in the

back room." He smiles. "Not for public viewing. Some of the specimens represent my faulty diagnoses."

Reflectively he turns the skull in his hands and offers a brief lecture. "It's a sea otter, of course. An animal's skull is a capsule history of its evolution. Eye sockets large, perhaps to accommodate large eyes for better vision underwater, or perhaps to accommodate a padding of fat to protect the eyes during deep dives. Bones with a purplish cast, surely caused by a diet rich in purple sea urchins. Teeth quite human, and the same number as in man, thirty-two . . . adapted more to grinding than to biting . . . canines shorter than in most carnivores. One can deduce that the otter rarely uses its canines for fighting and rarely grabs fish with its teeth. I have the skull of a very old otter whose teeth were worn nearly flat to the gums by chewing coarse foods, some of them doubtless covered with abrasive sand."

He continues. "I've read that it's possible to tell a sea otter's age from its teeth, though it has to be done in a lab. There's a microscopic growth layer on the root representing each year of life—a sort of register to time. When Alaskans were killing otters for the fur market they looked at all the skinned carcasses and found a few from animals over twenty years old."

The beaches of central California are visited in summer by several million tourists from many parts of the world. They shriek in laughter, call for their young, and play radios night and day. They charter fishing boats to crisscross the waters where Barney feeds. The fumes of gasoline, suntan lotion, and hamburgers-with-onions drift to his sensitive nostrils. But he is conditioned from birth to accept these transgressions of his habitat. One imagines that his basic lifestyle is little changed from that of the otters who lived here before the coming of man.

His typical "plan-of-the-day" in summer runs about as follows:

loafing	60 percent
feeding	25
grooming self	5
swimming	5
interacting with companions	5

If one combines feeding, grooming, and swimming as "work," the total represents about five of the fifteen day-light hours. And when one remembers that Barney is also active at night, he is surely far busier than the average land carnivore of his size. Biologists once fitted a free-ranging otter with a radio-tracking collar. Every time he dived, the strength of its signal faded. They learned that he spent nearly as much time feeding at night as during the day.

No one has yet determined how well a sea otter can see at night or in the depths of the sea. An otter's vision is presumably poorer than that of the harbor seal, its far more ancient and more specialized cousin. In clear water a seal can detect a moving object by moonlight at a depth of 115 feet and by full sunlight at 1,528 feet!

On the flooding tide of this particular afternoon, Barney combines feeding with fun. He has positioned him-self near shore in the gray-green reflection of the Big Sur Forest. He faces a half-submerged rock, the seaward side of which is a cliff girdled at the half-tide mark by blue mussels. He grabs a cluster between his paws as a rising swell picks up his rear end but leaves him holding fast to the mussels. His feet and tail sway with the water like ribbons of kelp. The brown, thready attachments of the mussels begin to separate from the rock. As the swell falls he rides it down, carrying the cluster with him. Again and again he thus uses the force of the sea in combination with his own body weight to gather a favorite food.

By evening he is at rest. The smoky sun, which had painted the Big Sur Forest with ruddy light, has set. Suddenly the air above him is filled with birds looping and dipping in quick flight. He lifts his head as he always does when danger portends, see only swallows and a scattering of Bonaparte gulls, and resumes his nap.

The birds are feeding on termites, which, emerging by the millions from decaying timber, are being wafted on the offshore breeze to certain death by drowning or predation. This late August night is their night to find mates. They flutter on loose, transparent wings and are easy prey for the birds.

When daylight comes the termites are gone. A dozen swallows, perched shoulder to shoulder on the twigs of a drifting snag, chatter in competition as though reliving the feast of the night before.

Meanwhile, two volunteers of The Network have been studying a female otter who often comes ashore at Century Point. She hauls out on a slanting rock covered with a turf of red algae and green surfgrass. She frequently shares the rock with a spotted seal who tolerates *her* presence, but plunges into the sea whenever a person comes in sight. The volunteers—a retired physician and his wife—are surprised this morning to see the otter deliberately turn her head across her shoulder, stick out her tongue, and lap seawater from a shallow basin in the rock.

"It's hard to believe," says the doctor. "If the otter were a patient I'd label her behavior *mariposia,* a craving for salt water. . . . On second thought, sea otters must get a good deal of salt in their diet. Shellfish tissues have a salt content roughly the same as in seawater, around 3.5 percent." He studies the animal thoughtfully, then lowers his glasses.

"We could suggest to someone at the Marine Station that a study be made of the salt tolerance of sea otters, using our friend here as subject. We know that she, at least, drinks seawater. And the experiment would be harmless."

His idea bears fruit. A task force including the physician, his son-in-law, and Finn Peterson capture the otter in a rope cargo-net while she sleeps. They carry her by truck to the Marine Station, which is equipped with salt-water pools. The otter tames quickly, accepting fresh squid from her keeper's hands after only three days in captivity.

At this stage a physiologist begins to study the rate of salt clearance from her blood. After injecting into her veins a harmless amount of radioative tracer (tritium), he monitors the appearance and disappearance of the tracer in her urine and droppings. Knowing the original salt content of her food and of the water in her pool, he concludes that she can handle the equivalent of two quarts of seawater a day in addition to the salt that she takes in with her food.

"The sea otter has a huge kidney," he says, "which explains how it can handle levels of seawater that would poison a land mammal. I've autopsied several sea otters found dead on the beach. Body weight for weight, the kidney is twice the size of a land otter's.

"People tend to blame salt for high blood pressure, and they have some reason. But salt—pure sodium chloride—is not itself a poison. On certain islands off the coast of Scotland, domestic sheep live the year round on seaweeds. During summer they graze on fresh weeds in the tidal zone and during winter on weeds torn loose by storms and washed ashore."

September

The September meeting of Otters Alive is being held at Life Arena to consider the question, "Should California Sea Otters Be Used for Research and Education?" A young man rises to his feet.

"As curator of Life Arena I should explain that we run a public aquarium. We're in business to offer entertainment and education, and to provide facilities for research and for the breeding of endangered species. I don't need to defend our interest in entertainment. Millions of Americans visit aquariums, zoos, and circuses purely for fun. In Life Arena, though, we put education above entertainment. When one of our killer whales is performing in front of an audience, we describe what it's doing in terms of what a *wild* whale would be trying to accomplish by the same behavior.

"The sea otter is a good example of an animal whose habits, until the first captives were exhibited in the 1950s, were known only to fur hunters. Today I suppose that, where one person has seen otters swimming underwater in the wild, a million have seen them swimming underwater in aquariums.

"As far as research is concerned, our biologists have contributed on many occasions to scientific knowledge. By feeding sugar to sea lion pups and later analyzing their blood, we learned that sea lions can't tolerate disaccharides, so it's no good trying to save an orphan pup by feeding it canned milk, which contains lactose.

"From watching our Ganges River porpoise, we learned that it habitually feeds on its side rather than on its belly, meanwhile waving its head from side to side and emitting sound pulses. River porpoises are blind from birth. Thus they evidently use sound alone to locate food in the murky waters of their native India.

"The scientific community knew almost nothing about

the birth and early parental care of cetaceans until the first captives became available for study in the 1940s.

"Some of you will remember our experiment with Kalan, the male sea otter whose vision we tested. By challenging him with underwater targets—white disks—of various sizes, and by rewarding him when he answered correctly, we found that he could see nearly as well when submerged as when on the surface. This ability, I think, represents an evolutionary compromise between the otter's need to find its way in two different media—water and air."

The speaker hesitates. "Knowing that sea otters often swallow stones, I once suggested that we test an otter's color vision by scattering marbles of different colors in his pool. If he could in fact discriminate, he would tell us so by selecting his favorite colors. We would later count the reds, blues, and so on in his droppings. Our staff veterinarian vetoed my suggestion. '*No way!*' he said. 'Too risky.'

"I could give other examples of the contributions of aquarium animals to research, but I'll stop now and entertain your questions."

The chairperson responds. "I would guess that most of us in Otters Alive are agreeable to the use of sea otters in research when the objective is worthwhile, when the animals are willing subjects, and when they aren't subjected to pain or fright. But some of us do object to *killing* them for science. Our reasons vary from reluctance to depleting California's small and threatened otter population to deep personal feelings against the taking of life. Our reasons range, you might say, from the practical to the religious."

"I understand you," says the curator of Life Arena, "and I'm a member of Otters Alive and am sympathetic with your goals. I suppose all that any one of us can do is to present his or her personal opinions and hope to be persuasive."

Penny Moreno, her chin high, rises to her feet. "The otters were here for thousands of years in natural balance with their food supply and with their enemies—sharks, killer whales, and Indians. I'm not a biologist, but it seems to me that management of otters will be easy when we learn to manage *people*. Within the last hundred years people killed nearly all the California otters. Then the foods that the otters had fed on—the urchins, abalones, clams, crabs, and lobsters—began to increase. But shellfish abundance couldn't hold up when Californians by the millions began to pour onto the beaches."

Her eyes flash. "If we're going to survive as cultured, imaginative people, we simply *must* agree to share our shores with wildlife of all kinds. We as men and women, not the otters, are on trial."

"Hear! Hear!" cries Finn Peterson, arriving late, briefcase in hand. "And we must get used to protecting the otter's environment, too. Enhydra can live only in coastal waters, and these are the very waters we're polluting with our wastes and poisons. The question is, will we learn to respect the seas before we've finished off the otters? It will be a close thing.

"More than three billion gallons of industrial and domestic wastes pour into California's marine water *daily*. Although they're relatively free from harmful bacteria, they do contain poisons—especially heavy metals and organic compounds—that aren't removed by ordinary sewage treatment. Cadmium, silver, mercury, organochlorine pesticides, and PCBs have been identified in the tissues of California otters. Some of these poisons, in animals of other species, are known to be linked with abortion and runty offspring.

"I seem to be lecturing again," says Finn wryly. "Let me end on a lighter note." He fishes a piece of paper from his pocket. "It's something I clipped from a wildlife magazine." He reads.

" 'Most of the troubles we are causing to other species are not . . . the spectacular ones. They are private pains which we can only know about through the occasional accidental discovery.' "

A respected scientific consultant to Otters Alive is the final speaker. She says, "Whatever we may have in mind for the good of the otters, they seem to be regulating their own numbers nicely. Having adaptable food requirements, they can satisfy their hunger in waters long-since fished out by man. No food species has ever disappeared where they roam.

"As a scientist, I see the otter as a natural member of the inshore community. Its population in California should be restricted neither geographically nor in numbers before it reaches a natural limit, and before the profound influences of that limit-population on the marine environment shall have been studied."

On one of the last sleepy days of summer, Barney and Paleface hear a splashing as of heavy bodies tumbling in the water and they periscope to see who's coming. They spot three otters—an adult male and a female with her large pup. Barney gets a whiff of the female and is sexually aroused. The strangers, oblivious to Barney and his mate, roll and frolic and fondle one another with their forepaws.

When at intervals the female becomes submissive, lying on her back with paws extended, the pup tries to climb aboard but is repulsed. After ten minutes of rough play, during which time she must alternately push the clinging, crying pup from her head and the clinging, snorting male from her rear, she rolls herself in a circle, uncoils like a steel spring, and shoots to the depths of the sea.

If one could see the fetal pup now growing in Paleface's womb, it would seem a stretched-out kitten with a body too large for its head. It is four inches long, naked, pink, and already displaying the unique features that mark it as a sea otter. Its tiny limbs have advanced beyond the generalized mammalian stage to become paws and flippers. They can move only feebly; Paleface has yet to "feel life."

Barney and Paleface, loitering in Whalers Cove this morning, are restive. At intervals during the previous night, full moonlight had struck their faces from a most peculiar source—the land! The source was, in fact, a powerful searchlight manned by two men from The Network who had stationed themselves on a cliff. They had been estimating the percentage of time spent by otters diving for food during hours of darkness.

As the men were stowing their gear at dawn they had glanced down at the edge of the cove and had seen an otter behaving strangely. It was trying to feed in water barely deep enough to cover its body.

"A half-grown female," one had said, lowering his binoculars. "Looks as if she's been smashed by a propeller. Her nose is cut and swollen and she favors her left arm as if it were broken."

"Should we try to take her to a vet?"

"I'd vote against it. Let's come back later and try to feed her. Chances are she'll recover—if she's ever going to —at the place where she feels at home."

So, with the help of friends in The Network they arrange for a daylight watch over the cripple. They toss fresh squid to her. They chase away dogs, and a man who is trying to hit her with stones. They feel encouraged when she crawls from the water and nervously takes food from their hands. When, however, one of them touches her head she instantly draws back.

"It's queer, you know," he comments now. "An otter doesn't like to be touched unless it's in the water. There it can become quite tame. When I lived at Point of the Dunes I made friends with a wild otter who became almost a pet. I would dive with him and he would snuggle his face under my chin and let me scratch him. He would grab my arm and pull himself up to sniff my mouth. Once we swam side by side for a good half-mile. I often tried to coax him onto the beach, but he wouldn't leave the water."

"I suppose," offers the second man, "that an otter knows instinctively that his life depends on his fur. He's nervous about anything that could disturb its arrangement or lay. A cat's the same way. Did you ever rub a cat's fur the wrong way? Plenty of excitement."

On the morning of the seventh day the two men find the little female lying on the sand trembling violently, opening and closing her jaws. She has voided black tarry droppings. They watch her soberly until she ceases to move. . . .

There comes a time to every animal when a critical tissue in its body, a critical cog in the living machine, fails to work. The machine cannot then be restarted. Although nature has great recuperative powers, she will stand for only so much abuse.

Amelia, from her office at the Marine Station, phones Penny in excitement. "There's a big male otter feeding alongside the station pier. I'm told that he's been here for several days. If you want to get photographs, now's your chance."

Penny replies, "Great! I'll be over in ten minutes. Finn is here with me; I'll bring him along."

When they join Amelia at the shoreside end of the pier, she unlocks a gate. "The public's not allowed on the pier. Too many sensitive instruments. The otter seems to have learned that he won't be bothered here. Anyway, he's completely unafraid."

Fast-moving clouds change the summer sea from blue to gray and blue again. Suddenly there's a *poof!* and an otter surfaces ten feet from the pier. He sees human figures against the sky and retreats a short distance, then begins to groom.

"Will we spook him if we move?" whispers Penny.

"No problem," Amelia answers. "He's used to people."

Penny maneuvers her camera into position and begins to snap frame after frame at one-thousandth of a second. When at last the animal disappears on business elsewhere in the bay, the three adventurers retire to Amelia's office for coffee.

"You know, of course," says Finn, "that a sea otter has a tough time keeping his fur clean and fluffy. It's his insulation against cold. Every time our friend out there patted his fur you saw small explosions of bubbles. Among the fur hairs there must be zillions of tiny air bubbles providing a sort of air-foam insulation."

"I've read that there's a dry air layer next to the skin," says Amelia.

"Maybe so; it would be hard to prove." He pauses. "It's probably a mixture of air bubbles, seawater, and sweat . . . like inside a diver's wet suit. Several years ago the agency financed a study of sea otter fur and skin. The fellow who did the work reported that the pelage has three hundred thousand fur hairs in an area the size of a penny."

"Three hundred thousand! How did he count them? Stay up nights?"

"He sheared a piece of skin and shaved it, leaving the cut hairs standing like stumps on a logged-off forest. He dried the piece and pressed it into a warm plastic film, getting a transparent cast, which he could then study under a microscope. He counted hairs on a known-area portion of the cast."

Penny rubs her chin thoughtfully. "But don't otters molt? Don't they shed their old coats for new ones in the spring, like dogs and cats?"

"You're wondering how they can afford to molt when they're bathed the year around in cold water. If they *were* to lose their fur in great wads, like dogs and cats, you'd think they'd die of pneumonia.

"As a matter of fact," Finn continues, "a biologist who once studied a sea otter in an aquarium pool answered your question. He put a wire screen over the outlet to the pool to catch any loose hairs shed into the water. At intervals he weighed the accumulation of hairs on the screen. He found that an otter sheds gradually throughout the entire year, though most rapidly in early summer."

"Is it possible," asks Penny, "that an otter has a racial memory of a time long ago when it lived on land and shed every summer?"

Finn smiles. "Could be. The important point, though, is that it sheds *gradually*. Evolution continuously favored those individuals who were reluctant to let go of their fur. Those were the ones who kept warmest at all seasons and the ones who survived to carry on their bloodline."

Finn stares through the open window at the sea. "We feel that otters have a tough time here in sunny California, but imagine what they must face in the Bering Sea—exposed in winter darkness to floating ice and below-freezing gales of wind. Their ability to adapt to cold seems as remarkable as the ability of desert animals to survive air temperatures that may rise above 120 degrees.

"As old Aristotle put it, 'In all things of nature there's something of the marvelous.' "

October

The first storm of autumn has come and gone, leaving a rain-beaten shore and a moving sea. The leathery forests of kelp are disarrayed. Plants uprooted by waves and currents are drifting against one another and against plants still rooted, knotting the living and the dead into huge balls like tangled yarn. Although one might suppose that nature's annual thinning of the kelp would harm the otters by reducing their shelter and support, it does not. Most pups, in fact, are born in midwinter—December through March—when the kelp is thinnest. Why this should be so is unknown.

Barney and Paleface have been feeding and playing together often during summer and fall. The bond between them is stronger than mere acquaintance, though weaker than the pair bond that holds, for example, the opposite sexes of wolf or of fox together beyond the simple needs of fertilization. Indeed, an observer might conclude, after seeing Barney repeatedly steal food from his mate, that he was hovering beside her for a purely selfish reason—to take advantage of her slower movements during pregnancy!

Now in a thin luminous fog before sunrise the two are feeding along a granite coast. Paleface has developed a

craving for fish, a food she rarely takes. Planing this way and that, she searches the bottom for sand dabs—little five-inch flatfish. She carries each to the top where, balancing her body in the surge and ebb of the sea, she delicately holds the fish between her forepaws, tears it apart with her teeth, discards the head and viscera, and swallows the flesh and bones.

Barney feeds with her for a while, then turns to hunting for abalones. He does not dislike fish, but he knows that some carry sharp spines that can pierce sensitive lips, while now and again one carries a barbed hook still attached to a length of fishline. So he dives to deeper water and swims on a winding course among bedrock shadows. Without warning, the water blackens and he tastes the peculiar fishy-musky flavor of octopus ink. Blinded, he jerks aside at the moment when the octopus also moves; the two collide. He bites deeply in the rubbery mass; his fangs reach the creature's brain before it can bring its powerful beak into action. Its yard-long tentacles writhe like earthworms cut by a spade. Barney leaps toward the open air, his face hidden by the body of the beast. On top, he blows explosively and begins to bite the tentacles one by one. Some fall from his mouth and twist out of reach, while others fasten their suction cups against the roof of his mouth, to be picked off by his claws. Later, his belly crammed with one of the sweetest foods of the sea, he rests. What a fortunate encounter!

He floats belly-up as usual, chin touching his chest. That he can hold his neck in the shape of a C for long periods is a demonstration of animal adaptation. There would have been a time long ago when a "proto sea otter" would have hauled out on the beach to devour its briny catches. Millennia later, a "halfway sea otter" would have eaten some of its meals at sea, holding its head and paws above the water. Today's sea otter is fully aquatic, using its body as a table on which to prepare and eat its food. The evolution of the sea otter's feeding posture was fundamentally the evolution of neck bones, muscles, and tendons —all of these being anatomical structures that changed by minutely small steps over millions of years.

Paleface moves to a kelp thicket to employ a new feeding strategy. After diving for a small rock and bringing it almost to the surface, she begins to burrow like a mole beneath the tangled foliage. Much as a person would pick hazelnuts, she picks off the little turban shells that graze in countless numbers on the fronds. When she has gathered ten or a dozen she emerges, dripping with weed, and cracks them singly against the rock held as anvil on her chest.

Near shore where the rhythmic press and release of the water moves like a song across her body, she finds a bed of mussels nearly smothered by ribbon kelp. Good! She grabs a kelp stem between her strong paws and lunges toward the surface. The whole plant, still holding fast to a cluster of mussels, comes with her. Floating on the surface with the kelp draped across her belly, she picks the mollusks one by one as a seamstress might pull pins from the hem of a dress. She crunches them in her teeth and swallows them, shell-bits and all.

Barney wakens and arches his back. He dives to a submarine prairie of surfgrass to nose among its green blades for abalones. Having found one, he rises to the surface and tears at its flesh as though he were starving. There always seems a quality of haste in the feeding of an otter, a quality more frantic than the circumstances would seem to warrant. Barney's eardrums are pounded by the steady roar of breaking surf, his body is rudely shoved by each incoming wave, and his open mouth is as often full of water as of air. No matter. He is wholly composed, for he is wholly a beast of the sea. When he has finished the abalone meat, he holds the shell between his paws and, oriental-fashion, tips the last drops of liquid into his mouth.

Now he bathes in a pool of whirling foam, moving his paws and his flipperlike feet simultaneously, though not in synchrony. While the paws perform one scrubbing routine the flippers perform another, as two actors playing horse inside a skin might move if head and rump were not communicating. Barney completes his toilet by "slicking down" —scooting full-speed ahead while rubbing first his right ear and then his left on the water.

As the rising sun colors the fog, Barney and Paleface move toward the shore. They begin to play—she as actively as her condition will permit. They compete in trying to nip one another and dart away before being nipped in return. Shoulder to shoulder they engage in pushing contests. Barney often pauses to shake savagely a bit of seaweed held between his teeth.

One would like to think that sea otter play goes beyond simply producing a strong, alert, well-coordinated individual. One would like to think that it generates amusement. Indeed, one would like to think that it makes for a happier individual. Sea otters *seem* to enjoy playing with one another and with pebbles or pieces of kelp, while land otters *seem* to enjoy sliding down snowbanks and chutes of slippery clay. Is it any less objective, less scientific, to grant these animals "feelings" than to suppose they are only machines of flesh and blood?

Now Barney leads Paleface toward the beach between massive granites that loom above their heads like ill-matched monsters. The two otters bobsled on a breaking wave over the crowns of a sea-palm grove. These tough, slippery plants have learned to survive by falling flat and rising with each incoming roller. With Barney still ahead, they enter a corridor among the rocks and climb a steep beach of rounded pebbles. They hear the little chorus as each new wave pushes the pebbles up the slope . . . then lets them roll down, chuckling. Barney picks a path among mounds of broken kelp torn loose by the storm. He walks with a clumsy, rolling, sailor's gait, his back arched and his belly dragging the ground, his hind feet slapping. Near the high-tide mark where the seaweed drift is rotting, black flies rise in swarms so thick as to veil the smelly piles in which they were hatched. Strewn here and there are pale jellyfishes and bright red fragments of cow's-tongue seaweed.

The two reach a sandy clearing. Barney rears awkwardly on his hind legs and searches carefully for enemies. He sees sandpipers and blackbirds pecking at kelp, cormorants drying their outstretched wings, and white egrets passing in file offshore; he shakes his fur and drops to the sand beside his mate. The two fall asleep, touched by the warming rays of the sun.

Not far away, an abandoned Spanish rancho, an old grazing land, slopes down to a sea-pounded shore. Finn had seen the land from a hilltop and had later scouted out the faint Indian track that meanders through a thicket from the coastal highway to the beach. First trod by the feet of deer and barefoot men, the track was later deepened by the hooves of cattle and ponies.

Today Finn and Penny have come early to this hidden place to enjoy a full day together, she to sketch and he to photograph and both to talk about life. In pushing their way through the chaparral they had startled into flight cottontail rabbits and ground squirrels, and a slow-moving lizard chilled by the night.

Now from a distant bush a white-crowned sparrow, low in spirit at this time of year, gives voice over and again to the same sweet, unfinished song. The ocean waves at their feet begin to speak in tones and overtones that carry through and beyond the range of the human ear. They whisper for long minutes . . . then boom in fury when they find themselves trapped in some rocky cave. The air is strong in seashore smells, primordial smells faintly sulfurous with the fumes of decaying weed.

"I hardly know where to start," says Finn, taking his camera from his pack.

"One bit of beauty at a time is my motto," Penny answers. "Let's begin with him," aiming her pencil at a solitary gull floating on a tidal pool in the kelp. The gull rises and falls and whirls in slow motion with the pull of unseen currents. Has he been rejected by the flock? Has he sought escape from some pain he cannot ease? He goes around, and around, and around.

Penny's fingers move lightly to capture the outlines of the lonely bird and his backdrop of weathered rocks. At the surf line the rocks are green with algae . . . higher, barren and gray . . . still higher, decorated with rosettes of silver stonecrop . . . and on the skyline, fringed with coyote bush in golden bloom.

Finn photographs the gull in various poses and turns to a heron standing on the water like a biblical figure. Through glasses, Finn sees that the bird is, in fact, supported by a bit of driftwood, waterlogged and barely afloat.

Penny stares hypnotized at a pool below her and speaks to herself. "It's an abstraction. It could never be put on paper or film . . . danced, maybe, or sung." Around the margin of the pool the waters are streaked with blue, turquoise, and white. At its center, sheets of yellow foam twist slowly, bumping against mats of chestnut-brown kelp.

Finn returns to her side. "Continents." He points to the sheets of foam. "Maps of imaginary worlds; lands that never were nor will be."

Her eyes soften. "You're quite the poet today. I have hope for you."

A sudden change at the edge of the sky draws their attention to a line of pelicans winging heavily along the coast to some rendezvous best known to their leader. As they pass, a sleeper wave still charged with the power of the recent storm explodes on the reef below them, silhouetting for a flash in time the form of a large gray bird. A willet, Finn decides before he and the bird are hidden in a cloud of salty mist. He wipes his eyes in time to see another line of birds—cormorants flying so near the sea that they disappear and reappear in the troughs of the rollers.

As Penny and Finn pick their way cautiously along the edge of a low cliff, they pause to watch an otter in the act of bodysurfing. Outlined in green, it hangs on the crest of a breaking swell until, at the very last moment, it ducks through and behind . . . to be lost in a welter of white.

"I think," says Penny, "he's using the surf as a laundry. Self-service. He's saving energy by not having to scrub himself."

Finn regards her. "That's one reason I like you. You're inventive. I'll buy the idea."

The two move closer to the edge of the cliff. "A battle-field," she says as she points down to hundreds of brown cylindrical floats torn loose from kelps and later thrown upon the sand. "Soldiers of one uniform, they lie where they fell."

He contributes. "They gave all for a cause they were never to understand."

Now the pair reach a vantage point from where they can look down on a most unusual beach. They have seen it before. The beach is a tiny mollusk cemetery, a strand built of white broken shells washed up by the sea. Penny halts abruptly, finger to lips. She stares at two otters lying side by side on the beach. They are Barney and Paleface, asleep in the sun. They are upwind from the human scent; they are unafraid.

The man and the woman retreat on tiptoe to follow their communion with the sea and the sand and the bright October air, and with one another.

November

In the late 1700s, while colonial Americans were fighting their War of Independence, a Spanish priest on the coast of California was watching an Indian capture a sea otter. Padre Luis Sales was later, in 1794, to describe the hunter's method:

> [The Indian] has provided a club and a long cord with two hooks, and when he discovers an otter he draws near it. The otter ordinarily swims carrying its young ones, teaching them to paddle with their little paws. Seeing the canoe she dives under the water and leaves her young on the surface. The Indian comes up immediately and ties the cord to a leg of the little otter so that one hook lies close to the foot and the other about a span away. This done the Indian retires with his canoe, paying out the cord, and when a little way off jerks the cord so as to hurt the otter, and it cries out because of the pain. At its call the mother comes, and as she sees the Indian is far away, she approaches it, clasps it and tries to take it away, but since the Indian holds tightly to the cord she cannot. Then the big otter tries by kicking its feet to get the cord off its baby

121

and usually gets entangled with one of the hooks. Now that it is caught the Indian comes up in his canoe with a club in his hand, gives it a blow on the head, and it is his.

A cold, woolly fog lies unmoving below the Santa Lucia Cliffs. Barney and Paleface are preening in a group of about twenty otters, many of them pregnant or with small pups. They form a nursery group in a bight hardly five acres in extent, protected by a headland. The air all around them is filled with the cries of unseen birds and with the faint, damp, organic smells of a winter shore.

Barney yawns, wheels headfirst into the sea, and dives. Bubbles spring from his pelage in silvery trains. The long silky hairs of his throat whip like little flags in a breeze. Arms folded on breast, he kicks his way in easy strokes. He seems a robot—tireless, controlled remotely, programmed to hunt with supreme economy of motion. When he reaches the bottom he follows a winding course above a cobbled pavement, his whiskers only inches above the rocks, his trunk slanting upward and swaying in pace with the strokes of his feet. Often he seems to spiral inside himself.

Having found a red crab, he rises, snorting as he clears the surface. When he has eaten, he turns to Paleface and encourages her to play . . . he wasn't very hungry, anyway. The two submerge and begin a liquid dance, writing the letters *O* and *S* in the blue-green water, colliding and pushing apart, embracing now and then like people.

They stop . . . and he is immediately alert.

Barney's ear has caught an odd little sound. His mate's attention is already fixed on a third otter lying on a mat of seaweed fragments. The pair scull toward the source of the sound.

"*Ee-yuh!*" A cry loses itself in the fog. A female is about to give birth. She bends forward twice in a sort of sitting-up exercise . . . and now thrice. A shiny purplish ball appears between her legs. Three seconds later a wet newborn pup in its amniotic sac is sinking through the kelp. The mother takes it with her teeth and lifts it to her breast, tearing the sac in the process. Now she knows exactly what to do. She kneads the sopping little thing on the blanket of her breast as though it were a ball of dough. It coughs feebly and draws its first breath; the umbilical cord breaks, dark blood dribbles into the sea. The mother clasps the pup in her arms.

Through some telepathy of the wild, a gull knows of the event. He appears from nowhere and paddles toward the madonna and child, darting his yellow bill at particles in the water.

In the meantime, Paleface has been circling the birthplace with intense interest, impelled to take part in a drama that she feels must have been written for *her*. She wants a piece of the action. But when she tries to nose the newborn pup its mother clicks her teeth rapidly and growls.

Until dark of midnight the mother allows the pup to rest on her breast. Hour after hour she grooms it ritually with tongue and paws as though knowing that its greatest need in this awful time is love. At last she pushes it toward her lower abdomen and lets it find the pair of life-sustaining teats, which will be the center of its world for months to come. She continues to lick its tail and rear even as it sucks. Its woolly coat grows magically thicker as she blows and whips air into its pile.

An hour later, she is beginning to shiver from cold and fatigue, so she places the pup in a cradle of kelp matching the color of its golden wool. It bobs like a cork, weakly moving its head and paws. Now she cleans her own soiled fur by rolling in the water and "toweling" every inch of her body. When four hours have elapsed from the time of birth she again cradles the pup in her arms and closes her eyes. The small brown head and the large pale one, touching, drift on a flowing sea.

In the days to come she will offer to her pup a measure of devotion unequaled among nonprimate animals. Rarely will the pup be allowed to stray beyond the reach of her paws, while her pink tongue will learn every part of its little frame. Her vision will be blocked by its rump and tail. She will seldom have room on her chest for the anvil rock she uses for cracking shellfish. Time and again her snout and whiskers will squeeze seawater from the puppy's wool, then comb and tease the wool into fluff. Patiently, and at all hours of the night and day, she will serve as launching pad and ferryboat. She herself will eat little during the first week, and then mainly at night, reluctant to dive and leave the pup exposed. (In Alaska, at least, eagles regularly prey on floating pups.) While the pup is helpless it will depend on mother for guidance to her teats. Later it will take the initiative, screeching with hunger and rudely mouthing the source of the milk. Still later it will swim to her while she floats belly-up, rest its head and forepaws upon her, and nurse with its body docked at hers in T-formation. If mother and pup are caught away from land by a sudden storm, she may begin to swim on a long oval course, moving forcefully against the slop of the waves, then sailing downwind to give the pup a chance to rest and suckle.

> They throw the young ones into the water to teach them to swim, and when tired out they bring them to shore again and kiss them just like human beings. They toss the young out into the sea and with their paws catch them when tossed, like a ball; and with them they engage in all the delightful and gentle games that a fond mother can play with her children.

Thus wrote (in Latin) George Wilhelm Steller, the first person to bring notice of the sea otter to the scientific world, in 1742. His description, though highly colored, tells how deeply he was impressed by the mother otter's care. She does not, of course, "teach" the pup anything, nor does any wild animal teach its young. The young are self-taught through observation, imitation, and trial-and-error. Steller's notion that the mother plays catch with her pup was perhaps the wistful thought of a lonely man finding himself on a Siberian island, remote in miles and years from his own fireside and his own children.

By the time it is two weeks old the pup will be grooming itself awkwardly, swimming a body length or two on its belly, picking at fragments of seaweed and spitting them out, and squalling like a spoiled child when it feels neglected. At age two months it will be diving in shallow water, staying down for as long as half a minute. At age five or six months it will start to feed for itself on the ocean floor, while intermittently nursing and stealing shellfish foods from its mother's chest. Whenever she feels too greatly abused she will roll over to cover her teats. And at age eight months the pup will be independent, though still a pest. It will continue to hang around her, to haul its great sprawling body upon hers as though she were a float, and to protest shrilly when it sees her remating.

Even when, at last, the pup has severed all physical ties to its mother, it will remember her and will remember the place where it was born. This, at least, is what some biologists think. They suppose that sea otters, like killer whales and social species of seals, have a sense of site fidelity or philopatry—attachment to the fatherland. These biologists suppose that philopatry, and interfamily ties, and the bonding together of families in rafts, are interrelated facets of sea otter society.

During the millennia while the "proto sea otters" were learning to live in the ocean, what sort of dens on shore did they last occupy? In what rude chambers did they last seek protection from the wind and the rain? Those chambers were surely no more than huddling places among beach-drift to which the female could resort while giving birth and nursing. Her fur dripping with seawater and weed, she would return periodically to the beach and to her young one. A few weeks later, when the pup's coat had grown thick and warm, she would carry the pup in her teeth to the kelp beds, there to continue its nourishment and to let it begin its marine education.

And what is the trait called "affection" that mother otters, as well as most mother animals, are loosely said to display? It is not a term in biology textbooks where, rather, one reads about epimeletic, or care-giving, or nurturant behavior. No matter. Karl Kenyon wrote

> I watched a mother carrying a dead, watersoaked pup while she emerged from the water and rested on kelp-covered rocks. For nearly an hour she licked the water from the pup's pelage and groomed its fur with her forepaws. When it was fluffy and dry she went to sleep with it on her chest. How long a mother will attend a dead pup is not known but one was observed in which patches of skin and hair were slipping from the body, indicating that it had been dead for several days.

At some time during the long prehistory of our own race, love emerged. Back of love was tenderness and back of tenderness was a set of automatic behaviorisms that improved the likelihood that an infant would survive to breeding age and would remember genetically, as had its mother, to offer care to *its* young. The sea otter's behavior is forever fixed near the automaton level. Yet, as care-giving is measured, it is a climax example. If one wishes to call it affection, no harm is done.

By the end of November the volunteers of The Network who had elected to scour the California beaches for dead and dying otters make their first report. They offer it through Finn Peterson at the November meeting of Otters Alive.

"I am pleased," Finn says, glancing at his notes, "to report on postmortem studies of thirty-two otters. The studies were made by biologists of the Fishery Agency and by local physicians and veterinarians.

"Six of the cases were emaciated pups, presumed orphans, while one was a stillborn fetus weighing less than a pound. Eleven animals had died either of enteritis—inflammation of the gut—or ulcers. It was clear in some cases that malnutrition or parastic worms had aggravated the inflammation. To our surprise we found no skin parasites such as fleas, ticks, or lice on any carcass. Enhydra may be the only mammal on earth—whales and seals included—that isn't bothered by external pests."

"Eight otters, young and old, displayed deep cuts, which we laid to sharks, propellers, spears, or bites from other otters. One animal still carried in its body a shark tooth torn from the fish that made the attack. Death by gunshot was confirmed by radiography in two cases; there were metal fragments in the brain. One otter had a badly ulcerated jawbone from a broken tooth. We think the infection handicapped his feeding and led to starvation. One female whose fur was heavily matted with crude oil had evidently died in a matter of days. I examined her myself. Her lungs were a classic picture of pneumonia."

Amelia Caring raises her hand. "I've been talking with fish experts at the Marine Station, and they agree that wounded otters are more likely the victims of sharks than of propellers. They say that great white sharks are here in our waters in all months of the year."

Finn hesitates. "I really don't know . . . but I'll try to think of a research design that would show the relative importance of sharks and propellers. You might suppose that sea otters, living as they do in a narrow zone along shore, would be pushovers for sharks and would quickly be exterminated. Not so. It's my own personal theory that sharks don't *like* otters. A shark will hit one, discover that the thick, loose, furry coat gets fouled up in its teeth, and will move on to hunt something that's easier to chew. My theory is supported by two facts: we find a good many otter bodies uneaten but bearing shark tooth marks, or actual teeth, and we've never yet found otter remains in a shark's stomach."

Finn resumes his report on stranded specimens. "The body evidence from one old female was that she died trying to give birth to twins. The first pup had been delivered; its broken cord was still hanging from the placenta. The second was lodged in the birth canal in breech presentation. I didn't see this case, but I read about it and saw photos of the specimens. During Alaska's commercial killing of sea otters, the biologists up there found that one pregnant female in sixty was carrying twin fetuses. That's a slightly higher incidence than in humans, where it's about one in eighty.

"On the other hand," Finn continues, "no one has ever seen a mother otter nursing two pups, and it's my belief that she couldn't support two. They would demand too much of her time and energy. She simply couldn't stand the strain and would eventually reject one or the other. To carry the point further, there's no record of *any* marine mammal, be it sea otter, seal, dolphin, or whale, having raised twins to independency.

"As you know, the sea otter has only one pair of functional teats. It would be interesting to study a series of early embryos. I shouldn't be surprised if at some stage in its development the embryo had *more* than two teat buds—a flashback, so to speak, to a time when the ancestral otter nursed a litter of pups."

He reflects. "We're dealing here with a species that's precocious at birth. A newborn sea otter weighs ten to fifteen times as much as a newborn land otter—partly, of course because the adult sea otter is the heavier of the two, but mainly because the newborn sea otter must be very well-developed to stand the shock of being launched from its mother's body into cold water. A sea otter at birth is already wide-eyed, and is equipped with a full coat of wool and ten sharp teeth.

"I never cease to marvel at the perfection of nature's plans for living organisms. If a newborn human baby were correspondingly as heavy as a newborn otter, the baby would weigh about fifteen pounds! Yet the enormous sea otter pup is delivered in a matter of minutes. Its rapid birth is surely an accommodation to the hazards of the sea. If *you* were a pup being born on a kelp bed in a February rain, you'd want to receive mother's tender loving care as soon as possible.

"Incidentally, there's a curious spinoff from our salvage program. We routinely saved the penis bones of all males as proof of sex. One of the bones we found had been fractured early in life, but had healed at the point of the break; another had been fractured and was still in two parts. I suppose an otter is occasionally caught by a breaking wave and is slammed with great force against a sharp rock. *Ouch!*"

Finn removes his glasses and polishes them with the end of his necktie. "Some of you may be wondering why The Network should spend time examining an otter after it's dead. You may think that whatever's learned by studying a corpse can't be applied to prolonging the lives of the living. In answer I would say that necropsies are valuable mainly for what they tell us about *changes* from year to year in the spectrum of injuries and diseases that otters suffer. If the changes point to an increase in human-caused damage by, for example, bullets, spears, propellers, or oil, perhaps we humans can take action to reverse the trend."

He adds, "I would personally favor a law against carrying firearms on boats, although I realize that this is not a popular opinion among my fish-and-wildlife colleagues.

"If, however, changes in the disease spectrum point to increasing damage from *natural* causes, there's little we can do about it. We can't change the behavior of sharks or keep otters from biting one another. We can't keep them from surfing near rocks in heavy weather.

"There's a possible exception. When natural mortality is increasing as a result of malnutrition, the trend can theoretically be reversed by cutting back the number of otters. You already know the arguments pro and con population control; I won't repeat them. They boil down to differences of opinion as to what's best in the long run for sea otters.

"Perhaps I have left the impression that an otter's death always has a single cause. Death, in fact, is usually the result of complications—of a chain of damaging events. Death is the enemy lurking outside the body's defenses; it can breach the walls at one place or at several. When an otter is being pounded day after day by rough weather, it may be unable to feed properly. It may lose weight and become fatally chilled. The cause of its death can then be entered in the autopsy book as storm, starvation, pneumonia, or drowning. One simply makes a choice."

Toward the end of November a cold front moves down from Alaska, bringing forty-degree weather to the central California coast. Whereas Barney in July had floated with his face, tail, and limbs upright in the air to cool them, he now does the same to warm them. But he keeps them as *dry* as possible, thus letting the sun's rays reach the skin through the sparsely haired surfaces of those appendages.

December

For a night and a day the wind has blown with gale force from the open sea, tearing the crests of the waves into smoke. As each sullen comber reaches the land it thunders against the rocks and throws yellow spume high in the air. This is one of the times when Barney finds the world hard to live in. Where he tosses in the lee of a headland he is partly protected from the fury of the storm, but only partly. An occasional gust hurls spray at his face like shot from a gun. He is in a holding pattern in a cove frequented by otters during rough weather—a place correspondingly depleted of shellfish foods.

But he must eat, come fair weather or foul. There is no relief for him through winter sleep, as for the black bear, nor through migration to a warmer clime, as for the Alaskan fur seal. So he dives to the bottom in water murky with debris and poorly lit by the storm-darkened sky. He hunts for snailfish, tadpole-shaped creatures about as long as a man's hand. Each carries under its chin a suction cup by means of which it clings to the bottom, shuffling along in search of worms, crustaceans, and small members of its own kind. When Barney finds one he slaps it, bites it, and

carries it to the surface, where he strips away its smooth brown skin and devours its flesh. He chews slowly. He might be thinking, *Not bad in the spring when full of eggs, but now only something to fill my belly.*

On the shore, gulls are tearing at the putty-colored body of a sea lion, probing its leaking wounds, bracing themselves lest they be scattered on the wind. Tonight raccoons and skunks will work on the body and, later, tidal sands will cover and uncover the skin and bones. By the end of winter only a dark stain and the faint sickly odor of death will remain.

Barney lost track of Paleface shortly before the storm broke. Three days ago she sensed the falling barometer, grew nervous, and started to swim slowly but directly toward Monterey Bay. She remembered a quiet place behind the Coast Guard breakwater where once she had loitered during a foraging trip. She is now in the eighth and final month of gestation, and is keenly aware of the small creature sharing her body. It has been making overtures daily more insistent, asking admittance to the company of otters. The jerking of its little spine and the thumping of its furry feet, together with new hormones flushing her own veins, disturb Paleface. Strangely, she wants to be alone—and now here she is inside the breakwater.

Her time comes in the black of a night pricked by the swaying lights of fishing vessels at anchor. It is a night when the ocean moans and groans and bellows without pausing to catch its breath.

Although the actual birth of a wild sea otter has yet to be recorded, one can imagine that Paleface's pup will emerge either headfirst or tailfirst, for in pregnant females examined by autopsy, half the fetuses lie one way and half the other. It being her first pup, it will perhaps be smaller than the average, say three pounds rather than four. It will sleep on her belly with its voice unused and its eyes closed for nearly an hour before it will waken to its new world. And for six hours on this dark night the bewildered mother will know her offspring only by feel, taste, and smell.

Barney, far to the south, is not aware that he is newly a father, nor would he have been aware had he been beside his mate when she offered their joint contribution to the long bloodline of Enhydra. Perhaps he would have been a little jealous. Who knows?

A ragged hole opens in the night sky. Barney and Paleface, miles apart, stare quickly at Venus where she brightens the heavens as she must have brightened them long ago when the proto sea otters crawled down from the land.

Meanwhile, members of Otters Alive are holding a Yuletide party in Monterey in celebration of their patron animal. Amelia has agreed to chair the festivities. She is impressive in a handwoven tunic from Africa, burnt-orange in color, alive with the old patterns of the land that gave rise to humankind.

"Friends," she says, opening the meeting. "We're here to pay respect to the fellow creature that has bound us together in a common struggle. It's a struggle that can never quite be won, for its objective is the preservation of life's diversity on a planet being overrun by people—an objective that seems to recede almost as fast as we approach it. Yet on we go, rewarded in the very act of struggling for what we think is right, what we think is natural in the deepest meaning of that word.

"I would say *thank you,* Enhydra. Thank you for helping us see that all life is an extension of ourselves . . . that your fortunes are our fortunes . . . that your future is our future. Thank you for showing us the web of life in which you fit so neatly and wondrously. Thank you for serving as a symbol of humanity's responsibility for our planet and its life. And thank you for helping us demonstrate to ourselves, the only race having the power to destroy you, that we can let you be as you are.

She smiles. "I didn't mean to be so serious. Let me call on our founder."

A gray-haired woman of distinctive bearing rises. She stands silent for a moment, then speaks about the animal whose unseen presence is felt by all in the room. Her words are clear.

"I wish to say only one thing. It is to thank all of you warmly for campaigning during the year to have the government declare the California otter colony a threatened population. And the government did so. This may seem a trivial act, but it's not. It promises that the colony will be free to grow at its own pace, and should it ever need to be controlled artificially, we common folk will have a voice in deciding how that control shall be applied. I feel that the government's action in this case is encouraging. It's a sign of a turnabout in public sentiment toward sea otters. From being thought of as simply a fishery resource, they have become a living national treasure."

Penny Moreno's eyes shine with pride, and Finn gives her a hug, when she learns that her artwork, *Sea Otter Heaven,* is to be reproduced as a poster and sold by Otters Alive for the benefit of the group. A silk-screen print now hanging behind the podium, it shows a yellow sun rising above the branches of a black cypress and pouring its warmth upon a bay where otters rest on beds of weed. A thousand diamonds sparkle along the track of the sun. It shows also the dim world below, where a solitary otter sinks into primeval green. Shoals of golden fish and silver squids caress his body. Within reach of his paws are vertical rocks crawling with the fattest of crabs, lobsters, and urchins. Curtains of kelp part soundlessly as he falls through and among them. Here indeed is "all earth forgot and all heaven around."

Amelia now reads the names of those in The Network who contributed during the year to the protection of sea otters and to knowledge of their ways in the sea. Loud applause. The animal doctor for Turtle Bay responds by reading a poem.

TO ENHYDRA

On the shore I find you and I know you.
When last we met in Mesozoic time we spoke a common
    tongue.

Traveler of wetness, deep roamer cutting the dark
You see the world through liquid eyes.
You hear the water-sounds.
You feel black stones at the bottom of the sea.
You rise dripping. You glisten.

From your floating bed you watch the stars
    how they stand against the night and
    how they fall.
You watch the moon tremble at the rim of the sky.

Though wild, you are lawful in your way.
Break otter law and you must die
    your seed unspent.
Stay so. Do not wish to be moral.

Wild beast, if you must go
    let it not be we who betray you.

The chairs are pushed to the wall, the punch bowl is
uncovered, and the walls of the old warehouse ring to the
rhythm of dancing feet.

The last day of the year sees Penny and Finn together
on the beach near Turtle Bay where the Indian mortar had
been found. They spent last night in sleeping bags on the
upper sands, sheltered by a tarp stretched on driftwood
above their heads. They did not hear the barking of coyotes

in the hills at moonrise, nor did they know that a fieldmouse
had traced the roundness of Penny's cheek as she slept.

Finn has now wandered off to look for a useful length
of mahogany or teak, or a fishing lure, or a bird skeleton,
or whatever prize the sea may have cast upon the beach
during the storm. To his way of thinking, a trip to the shore
is incomplete unless it yields a trophy he can carry home.

Penny is less concerned with material things, although
at the moment she is feeling her breakfast. *Sandwiches* . . .
It's the last time I'll let *him* plan the food . . . cold sausage
with ketchup . . . cold beans with ketchup. Then her lips
curve softly. *It's not all that important.* She opens her sketch-
book and returns to the project that has occupied her spare
time for twelve months—illustrating a diary of nature's
changes along the shore.

On a ledge above her head a cormorant in glossy black
preens himself with his bill. At her feet there lies a pavement of dove-gray rocks so smooth, so perfectly bedded, so
gracefully rimmed in their own shadows, that a quick little
shiver of delight races down her spine. Farther along, the
storm has exposed a sandstone terrace cratered, ridged,
and furrowed like the surface of some far-off planetary
moon. At one point on the stone a giant's eye returns her
gaze—a Cyclopean eye carved by whirling currents when
the Earth was young.

She poises her pencil to sketch a seaweed tapestry.
Shades of brown, green, and gold . . . stems reflecting the
sun and stems transilluminated . . . lines that are straight,
curved, or crinkled. Then she turns away, sighing, "I've got
books of weed sketches!"

She watches a wave rush through a channel to attack
her . . . it changes its mind and backs away in confusion.

She climbs a claystone bank in search of what she calls "my Henry Moores"—and others call clay babies or concretions —small figurines molded by nature . . . spheres, dumbbells, animals, people, and organic shapes that have no living counterparts outside the mind.

All at once in the slanting brilliance of the winter sun she feels wholly at peace with herself and with the world. She drops to the sand, turns flat on her back, and flings her arms wide. The rustling of sea grass and the wind's whine and the far-off crying of gulls resolve into one pure orchestral sound. She drowses. The tide creeps toward her feet.

. . . She wakens. Has a feather brushed her cheek? A faint, familiar, musky smell lingers in her nostrils. Wide-eyed, she turns her head toward the west and sees an otter sliding into the sea. He stops, raises one paw as if he were caught between two worlds, reluctant to enter the one or to leave the other. Then he is gone.

Penny will never see Barney again. After he left her he drifted south and took up residence among the broken rocks of a shore seldom visited by man. And, as far as anyone knows, he is still there now, gliding through the green canyons of the sea, knowing life as we humans can never know it, seeing shades of color we cannot see, vibrating to sounds we cannot hear. Chance had led him to cross the boundaries of experience between animalkind and humankind. When he returned to the ocean wilderness he left persons on the land who would always be richer in spirit for having tried to understand his ways, and for having tried to help him.

Index